Equine Supplements & Nutraceuticals

A GUIDE TO PEAK HEALTH AND PERFORMANCE

EQUINE SUPPLEMENTS & NUTRACEUTICALS

A GUIDE TO PEAK HEALTH AND PERFORMANCE

by Eleanor M. Kellon, VMD

04 03   8 7 6 5 4 3

For information address:
Breakthrough Publications
www.booksonhorses.com

Manufactured in the United States of America

Library of Congress Catalog Card Number: 98-74734

ISBN: 0-914327- 78 - X

Editor: Joan Vos MacDonald
Cover design by Greenboam & Company
Icons by Tricia Tannassy and Robert Greenboam
Typesetting and page layout by Peggy Hurley

Thanks to Michael Ball, DVM, Department of Large Animal Medicine, Cornell University; Jim Hamilton, DVM; and Lon Lewis, DVM; for reviewing this manuscript.

Abbreviated sources are given under each illustration. Full sources are as follows:

Maps on acid rain distribution are from the National Atmospheric Deposition Program, NRSP-3 National Trends Network (1998) NADP Program Office, Illinois State Water Survey, 2204 Griffith Drive, Champlain, IL 61820.

The chart on Nutrient Requirements and Balancing Rations for Horses, Table 4, by Dr. L.A. Lawrence, Virginia Cooperative Extension animal scientist, horses, is used with permission from Virginia Tech, Publication 406-473, July 1996.

The regional soil deficiency maps are used with permission from Dr. Ross M. Welch, plant physiologist, USDA-ARS U.S. Plant, Soil & Nutrition Laboratory, professor of Plant Nutrition, Department of Soil, Crops & Atmospheric Sciences, Cornell University. Feed analysis was done by TPC Labs, the Pillsbury Company, St. Paul, MN.

Photo Credits:
Cover photographs by Grant Heilman
Gemma Giannini, Grant Heilman Photography, p. 144
James M. Mejuto, p. ix and p. 96
Mark Newman, Tom Stack and Associates, p. 23
Dusty Perin, p. 37
Marilyn "Angel" Wynn, p. 1 and p. 216

Impreso por Imprelibros S.A.
Impreso en Colombia – Printed in Colombia

This book is dedicated to
my husband, Andy,
my most trusted and valued supporter,
advisor and critic.

CONTENTS

Acknowledgments

Introduction   ix

Chapter One   1   **Basic Nutrition**

About This Chapter
Various Feeds
Sweet Feeds and Complete Feeds
Factors Affecting Mineral Levels in Feeds

Chapter Two   23   **Mares and Growing Horses**

About This Chapter
How Various Feeds Measure Up

Chapter Three   37   **Nutrition A to Z**

About This Chapter
A to Z—Nutrients Recommended as Important by the National Research Council

Chapter Four   96   **A to Z Other**

About This Chapter
A to Z Other—Nutrients with No Established NRC Recommended Level of Intake

Chapter Five   144   **Nutrition and Performance**

About This Chapter
Supplements Used to Improve Performance

Chapter Six   174   **Relieving Health Problems Through Nutrition**

About This Chapter
Index of Health Problems Suggestions for Health Problems

Chapter Seven   216   **Consumer's Guide to Supplements**

About This Chapter
Determining Your Supplement Needs
Picking a Supplement
Choosing Your Basic Supplement

Glossary   225

Index   227

INTRODUCTION

The Thoroughbred has a uniquely high aerobic capacity, which influences the kind of supplements that may be given.

"With magnesium you only have to make the intake a tiny bit low to whack performance severely."

"CoQ$_{10}$ supplements improve performance without any change in exercise, maximum exercise capacity increased a whopping 28 percent."

"Even one bout of intense exercise to exhaustion can reduce muscle glutathione by 40 percent."

"Carnitine reduced post-exercise levels of lactate and pyruvite and significantly increased maximal work output."

"During the anaerobic phosphate trials, maximal power output increased by 17 percent."

These are a few quotes on how supplements enhance performance from *Optimum Sports Nutrition* by Dr. M. Colgan, Advanced Research Press.

This is just a tiny sampling of the solid, useful, scientifically proven information to be gained from just one 562-page sports nutrition book—if you are a human athlete.

Every day millions of people check the nutrition labels on cereal boxes to see how they measure up in supplying the RDA of vitamins and minerals. Millions more take a complete multiple vitamin/mineral supplement, in addition to their diet, providing all the day's minimal requirements in one dose. Still more take a wide variety of supplements and/or megadose nutrients to ward off aging, protect their cells from damage caused by environmental toxins, enhance their athletic performance or treat a wide variety of disorders—from failing eyesight to cancer.

There is absolutely no doubt that a correct program of supplementation can improve performance (often very dramatically), prevent injury/illness and speed recovery times after an intense effort. In every species where it has been tested—from rats to chickens to humans—the results are clear. When nutrients are used in this therapeutic sense, they are often referred to as nutraceuticals.

Nutritional causes for, and treatments of, disease in horses is not a new concept or an off-the-wall alternative therapy. Poor hoof quality is treated with biotin, methionine and zinc; tying-up treated and prevented with vitamin E and selenium; bleeders controlled with vitamin C and bioflavinoids; infertility in mares linked to problems with vitamin E, vitamin A, iodine and selenium; osteochondrosis dessicans (OCD) tied to inadequate calcium, copper, phosphorus and zinc.

Suboptimal nutrition will have deleterious effects on virtually every organ system and function of the body—reproduction, bones, joints, tendons, muscles, lungs, etc. Even the personality is influenced, for better or worse, by nutrition.

Yet many people, perhaps even many of the same people using vitamins themselves, believe all the horse needs to maintain peak health and performance, remain free of lameness problems and perhaps even break records, is timothy hay, oats, water and salt.

Even those who recognize, or at least suspect, that supplementation would be beneficial (if not necessary) have a very difficult time finding out what to supplement or how much to give.

Equine sports medicine is in its infancy—actually little more than an embryo. Interest in safe and legal methods to improve performance times is always high but information nearly impossible to come by.

Those who are waiting for scientific proof of effectiveness in horses are likely to have a very long wait. Horses are expensive research animals to keep, not to mention prone to all sorts of problems that could wreck the most carefully designed research study.

There is very little government or private money available to fund studies. Drug companies are not interested since there is no way to protect their profit potential using natural, nonprescription supplements. Companies manufacturing supplements are under no obligation to prove their ingredients and/or dosages are effective as long as they are careful not to make any medical claims (e.g., treating or preventing a disease or disorder). There is not much motivation to dump money into expensive controlled studies. Horse owners and trainers must rely instead on testimonials from users.

The basic reference bible on equine nutrition is the *Nutrient Requirements of Horses,* published by the National Research Council in Washington, D.C. This government agency publishes general feeding guidelines as well as recommended levels of several important vitamins and minerals. Recommendations from the NRC are MDRs—minimal daily requirements, defined as "the minimum amounts needed to sustain normal health, production and performance of horses."

In the introduction of the fifth edition, the authors freely admit the recommendations are not averages of all data available but are based on selected works of published research (not all published research), and extrapolations are freely made between types of horses. For example, if the only solid information concerning the nutritional needs of 11-month-old weanlings was from research done using ponies, that data will have been adjusted as the subcommittee saw fit to cover the needs of weanlings of all breeds at that age—from Belgians to Thoroughbreds—based on their "experience in applying the information to field situations." In other words, many of the recommendations may be little more than an educated guess.

Where solid, research-backed information does not exist at all (which is the case for many of the vitamins and minerals), recommendations are made by calculating how much of the vitamin or mineral in question would be consumed by an average horse on an average diet.

It is probably no coincidence that recommended vitamin and mineral intake in many instances happens to be virtually identical to what the horse would be getting if he were eating a diet of timothy hay and oats.

NRC guidelines are used by many manufacturers of supplements and all manufacturers of equine feeds. The problem is, following these recommendations will, at best, leave you with a horse in normal health (which translates as essentially free of any obvious signs of a severe disease) that should show normal (translation average) growth, reproduction and performance.

If you think of your horse as average and are satisfied to have him survive without glaring signs of deficiency diseases and putting out average performance, put this book down.

Remember that accepted as "normal" plagues for the "average" horse are such things as navicular disease and a wide range of degenerative arthritic conditions, poor reproductive performance, hoof quality problems, tendon and ligament injuries, muscle pain, back pain, premature aging of performance horses, hormonal imbalances, a host of skin problems, assorted allergies and lung problems and behavior/learning difficulties—all of which may have a nutritional component, often a very large nutritional component, and be very responsive to dietary adjustments and supplementation.

If you want more in terms of overall optimal health, prevention of disease/injury—and decreased recovery times when they do occur—plus peak performance, read on.

This book will attempt to fill in many of the gaps in available knowledge about supplements. It will look into what is known and unknown, in terms of dietary requirements and toxic levels of vitamins, minerals and other supplemental nutrients. It will investigate what is known about nutrition in other species and how that might be applied to the horse.

It will look at common health, lameness and performance problems and what the role of supplementation might be in correcting them.

Finally, some currently available supplements will be cataloged with a list of nutrients they contain and in what amounts.

The use of nutrition to maximize health and performance and to prevent and treat disease is highly effective and safer than conventional drugs.

It is a science backed by many of the world's best minds working on riddles of nutrition in many species. It is also an area that has been sadly neglected in the care and management of horses. The time has come to tap into this vast resource.

By the time you've read to the end, you will be armed to put your horse in the best possible state of general health, with a few methods for getting the best performance possible as well, without the risk of positive tests for illegal substances or harmful side effects. Maybe this book can also catch the interest of possible funding sources, so we can finally have access to horse-specific information in this potentially very profitable approach to maximizing performances.

FLOW CHART

This flow chart will give you an idea of how you might use this book to supplement your horse's diet.

1–NEEDS

Determine what your horse's needs are. Will your horse work only moderately hard or will he run in high speed endurance events? Supplement needs vary according to energy expenditure.

2–ANALYSIS

Analyze your horse's feed, using Chapter 1, Basic Nutrition. For example, if you have a horse that will be taking part in three-day races and is eating only timothy hay, his or her diet may be deficient in zinc.

3–SUPPLEMENTS

How important are missing supplements? If you check Chapter 3, Nutrition A to Z, you will find that zinc is important to immune function, healthy skin and may be of benefit in horses heavily exercised.

4–HEALTH

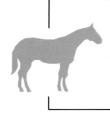

To consider any health problems that may impair performance, see Chapter 6, Relieving Health Problems Through Nutrition. For example, zinc may help with certain skin problems. Look up skin problems for suggested dosages. Be sure to note any recommendations about possible toxicity. Consult your veterinarian if you do not see improvement.

5–EVALUATIONS

If you conclude that a supplement might help, go to Chapter 7, Consumer's Guide to Supplements. This chapter explains how to evaluate different forms of supplements to see if you are using the most effective type.

CHAPTER 1

BASIC NUTRITION

Every diet of hay and unsupplemented grains requires the addition of free choice salt.

ABOUT THIS CHAPTER

This chapter is designed to show you how various diets measure up to the nutritional needs of horses at different levels of activity. The horse used in all calculations was a 1,100-pound 5-year-old (either sex).

The basic diets used consist of either timothy, alfalfa or mixed (half of each) hays only or the same hays fed with oats, using equal amounts of each, e.g., 100% alfalfa or 50% alfalfa and 50% oats. Nutrient content of the feeds was taken from NRC averages for midbloom hays, oats and grain. The half grain and half hay diet would be very unusual for many of these classes of horses. However, it

illustrates how the nutritional breakdown of the diet changes when you substitute grain for hay.

To get a completely accurate evaluation of your horse's diet you MUST have the actual grain and hay you are feeding analyzed by a laboratory.

The numbers used in creating these charts are the averages published by the National Research Council and do not necessarily reflect the recommendations given elsewhere in this book. They also may not accurately reflect the nutrient composition of hay in your specific area or for the same type of hay from year to year. Mineral, vitamin and protein contents of the same type of hay may vary widely depending upon the area where it was grown, the weather that year, fertilizers used, etc. It will be clear from these charts that simply feeding hay or hay and grain alone can lead to deficiencies, even for horses at maintenance and even by the NRC's very conservative guidelines.

To generate the charts, diets were first balanced for energy, using a reasonable estimate of how much the horse could be realistically expected to eat. This means that the first step was to calculate how much of the specific hay or hay and grain diet the horse would have to eat just to maintain his weight. An upper daily intake of feeds that a 1,100-pound horse could realistically be expected to eat on a consistent basis was set at 22 pounds for horses from maintenance through moderate work; 26 pounds for endurance/race/heavy work. This resulted in some of the hay-only diets being inadequate in energy (calories) for horses in work.

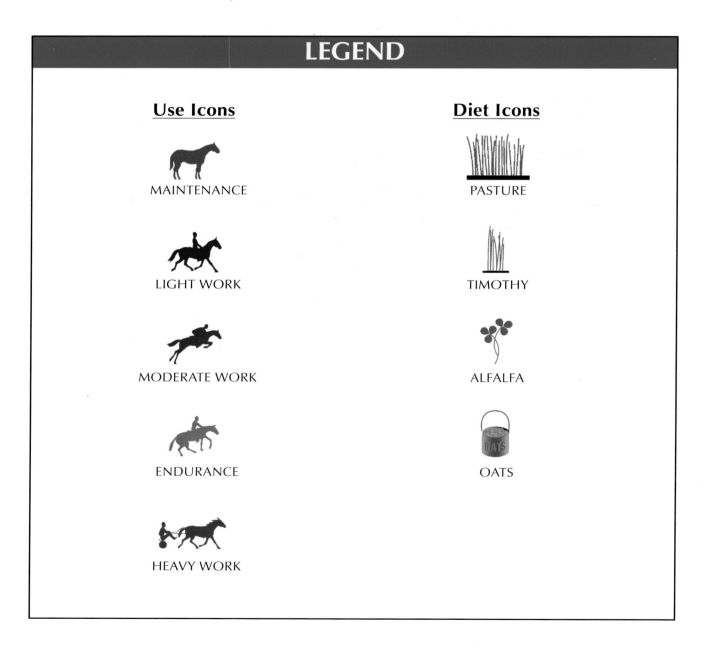

LEGEND

Use Icons

MAINTENANCE

LIGHT WORK

MODERATE WORK

ENDURANCE

HEAVY WORK

Diet Icons

PASTURE

TIMOTHY

ALFALFA

OATS

UNLIMITED PASTURE

DIET DISCUSSION

AMOUNT CONSUMED: 64 pounds/day

Pasture is obviously the most natural source of nutrients for the horse. Nothing puts the glow of health on a horse like access to healthy, young pastures. However, grasses have a very high water content compared to hays. This means that the horse must spend a considerable percentage of his time simply eating in order to take in enough calories and other nutrients. In fact, that is exactly what horses in the wild do—spend the vast majority of their time eating. Simply to maintain his body weight, a horse will need to consume approximately three times as much grass as hay. (This is only an estimate. Easier digestibility of fresh grass could lower the amount needed.) As the pasture ages and the quality of the grasses decline, they would need to eat even more. Wild horses are slim and trim not only because they might get more exercise but also because they simply cannot eat enough grass to put on and/or keep on significant amounts of body fat.

DEFICIENCY ANALYSIS

Legend:
- ■ (green) adequate
- ■ (yellow) marginal to inadequate
- ■ (red) inadequate
- □ (blank) level of this nutrient unknown

Activity	DE	CP	Ca	Ph	Mg	K	Na	C	Su	Fe	Mn	Cu	Z	Se	I	Co	VitA	VitD	VitE	Thia	Ribo	Lys
(horse)																						
(horse)																						
(horse)																						
(horse)																						

Abbreviations:

DE digestible energy	Ph phosphorus	Na sodium
CP crude protein	Mg magnesium	Cl chloride
Ca calcium	K potassium	Su sulfur

Fe iron	Z zinc	Co cobalt
Mn manganese	Se selenium	VitA vitamin A
Cu copper	I iodine	VitD vitamin D

VitE vitamin E	Lys lysine
Thia thiamine	
Ribo riboflavin	

TIMOTHY HAY ONLY

DIET DISCUSSION

Timothy hay has a correct ratio of calcium:phosphorus for horses, making it a favorite of many nutritionists. **Timothy is probably the ideal diet for horses at maintenance, by NRC Standards.** However, it is important to realize that timothy has helped to actually set the NRC standards (adult horses on timothy are free of glaringly obvious deficiency states so timothy must be an adequate feed for adult horses at maintenance). There is no NRC data available on the lysine content of timothy hay. This is not the same thing as saying it has no lysine; only that it has not been well studied. The same is true for selenium and timothy hay. As with all diets, free choice salt (NaCl) must be provided. Zinc, manganese and vitamin E supplements are also advisable. Timothy hay alone is not adequate for any level of work above light work.

If amounts are unknown, you should assume they are inadequate.

*Digestible energy inadequate at intake of 22.5 pounds/day for moderate or 26.1 pounds per day for endurance and heavy/racing.

DEFICIENCY ANALYSIS

Legend:
- ■ (green) adequate
- ■ (yellow) marginal to inadequate
- ■ (red) inadequate
- ☐ (blank) level of this nutrient unknown

Activity	DE	CP	Ca	Ph	Mg	K	Na	C	Su	Fe	Mn	Cu	Z	Se	I	Co	VitA	VitD	VitE	Thia	Ribo	Lys
horse (grazing)	green	green	green	green	green	green	green	green	green	green	green	green	green	blank	green	green	green	green	green	green	green	blank
horse (light work)	green	green	green	green	green	green	green	green	green	green	green	green	green	blank	green	green	green	green	green	green	green	blank
horse*	green	green	green	yellow	green	green	green	green	green	green	yellow	green	green	blank	green	green	green	yellow	green	green	green	blank
horse*	green	green	green	green	green	green	green	green	green	green	green	green	yellow	blank	green	green	green	yellow	green	green	green	blank
horse*	green	green	green	green	green	green	green	green	green	green	green	green	yellow	blank	green	green	green	green	green	green	green	blank

DE digestible energy   Ph phosphorus   Na sodium   Fe iron   Z zinc   Co cobalt   VitE vitamin E   Lys lysine
CP crude protein   Mg magnesium   Cl chloride   Mn manganese   Se selenium   VitA vitamin A   Thia thiamine
Ca calcium   K potassium   Su sulfur   Cu copper   I iodine   VitD vitamin D   Ribo riboflavin

4   *Equine Supplements & Nutraceuticals*

ALFALFA HAY

DIET DISCUSSION

Alfalfa has more calories and more protein than grass hays such as timothy. Since you feed less, horses at maintenance may not be satisfied to eat only the amount they require to hold their body weight. Alfalfa also contains far more calcium than phosphorus and can lead to bone/joint problems if this is not corrected (must have phosphorus at the minimum required level and calcium no higher than six times the minimum, but ideal is 1.5 to 2.0:1 calcium:phosphorus). As with all basic diets, free choice salt must be fed (NaCl).

Horses on alfalfa hay probably have borderline to deficient intake of zinc, manganese, selenium and vitamin E. High calcium may also interfere with the absorption of magnesium. Alfalfa is best used to boost the calcium, protein and caloric content of other diets, not as the base for a diet although horses in light work should do well on alfalfa alone, if missing minerals are supplemented.

* Amount needed to hold weight is less than the recommended total daily intake of dry matter. This means the horse will feel hungry and may have some digestive problems.
# Amount needed to maintain weight is greater than most horses could realistically be expected to consume in a day.

DEFICIENCY ANALYSIS

■ (green) adequate   ■ (yellow) marginal to inadequate   ■ (red) inadequate   ☐ (blank) level of this nutrient unknown

Activity	Nutriment DE	CP	Ca	Ph	Mg	K	Na	C	Su	Fe	Mn	Cu	Z	Se	I	Co	VitA	VitD	VitE	Thia	Ribo	Lys
*																						
#																						
#																						
#																						

DE digestible energy	Ph phosphorus	Na sodium	Fe iron	Z zinc	Co cobalt	VitE vitamin E	Lys lysine
CP crude protein	Mg magnesium	Cl chloride	Mn manganese	Se selenium	VitA vitamin A	Thia thiamine	
Ca calcium	K potassium	Su sulfur	Cu copper	I iodine	VitD vitamin D	Ribo riboflavin	

1/2 TIMOTHY HAY AND 1/2 OATS

DIET DISCUSSION

The main function of grains in the diet is to provide a more concentrated source of calories/energy for hardworking horses. As with all diets, a salt block must be provided. Vitamin E will be deficient. The drawback to grains is that as you increase their proportion in the diet, the mineral status of the horse suffers, as is clearly seen from the charts. Grains should not be used until a horse's activity level reaches moderate. At heavier work loads, grains must either be increased to over 50% of the diet or fat can be added as a more concentrated energy source (see Chapter 5, Nutrition and Performance).

* Amount needed to hold weight is less than the recommended total daily intake of dry matter. This means the horse will feel hungry and may have some digestive problems.

\# Amount needed to maintain weight is greater than most horses could be realistically expected to consume in a day.

DEFICIENCY ANALYSIS

■ (green) adequate   ■ (yellow) marginal to inadequate   ■ (red) inadequate   □ (blank) level of this nutrient unknown

Activity	Nutriment	DE	CP	Ca	Ph	Mg	K	Na	C	Su	Fe	Mn	Cu	Z	Se	I	Co	VitA	VitD	VitE	Thia	Ribo	Lys
*																							
*																							
#																							
#																							

DE digestible energy	Ph phosphorus	Na sodium
CP crude protein	Mg magnesium	Cl chloride
Ca calcium	K potassium	Su sulfur

Fe iron	Z zinc	Co cobalt	VitE vitamin E	Lys lysine
Mn manganese	Se selenium	VitA vitamin A	Thia thiamine
Cu copper	I iodine	VitD vitamin D	Ribo riboflavin

1/2 ALFALFA AND 1/2 OATS

DIET DISCUSSION

Grains in the diet provide a concentrated source of calories/energy, but as you increase their proportion in the diet, a horse may lose out on valuable minerals. Vitamin E will be deficient and a salt block should be provided.

Although the alfalfa and oats combination has fewer deficiencies than timothy and oats, supplementation is still required. Grains should not be used until a horse's activity level reaches moderate. At heavier work loads, grains must either be increased to over 50% of the diet or fat can be added as a more concentrated energy source (see Chapter 5, Nutrition and Performance).

* Amount needed to hold weight is less than the recommended total daily intake of dry matter. This means the horse will feel hungry and may have some digestive problems.

Insufficient to maintain weight at top expected intake.

DEFICIENCY ANALYSIS

■ (green) adequate   ■ (yellow) marginal to inadequate   ■ (red) inadequate   □ (blank) level of this nutrient unknown

Activity	Nutriment	DE	CP	Ca	Ph	Mg	K	Na	C	Su	Fe	Mn	Cu	Z	Se	I	Co	VitA	VitD	VitE	Thia	Ribo	Lys
*																							
*																							
*														(Cu white)									
#												(Mn white)											
#												(Mn white)											

DE digestible energy Ph phosphorus Na sodium Fe iron Z zinc Co cobalt VitE vitamin E Lys lysine
CP crude protein Mg magnesium Cl chloride Mn manganese Se selenium VitA vitamin A Thia thiamine
Ca calcium K potassium Su sulfur Cu copper I iodine VitD vitamin D Ribo riboflavin

1/2 MIXED HAY AND 1/2 OATS

DIET DISCUSSION

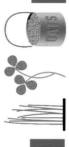

Although grains provide a more concentrated source of calories/energy for horses in work, increasing their proportion in the diet may decrease the presence of much needed minerals. This is clearly seen in the charts. Grains should not be used until a horse's activity level reaches moderate. At heavier work loads, grains must either be increased to over 50% of the diet or fat can be added as a more concentrated energy source (see Chapter 5, Nutrition and Performance). As with all diets, a salt block must be provided. Vitamin E will be deficient.

* Amount needed to hold weight is less than the recommended total daily intake of dry matter. This means the horse will feel hungry and may have some digestive problems.

\# Amount needed to maintain weight is greater than most horses could be realistically expected to consume in a day.

DEFICIENCY ANALYSIS

■ (green) adequate   ■ (yellow) marginal to inadequate   ■ (red) inadequate   □ (blank) level of this nutrient unknown

Nutriment

Activity	DE	CP	Ca	Ph	Mg	K	Na	C	Su	Fe	Mn	Cu	Z	Se	I	Co	VitA	VitD	VitE	Thia	Ribo	Lys
*																						
*												□										
#																						
#																						

DE digestible energy	Ph phosphorus	Na sodium	Fe iron	Z zinc	Co cobalt	VitE vitamin E	Lys lysine
CP crude protein	Mg magnesium	Cl chloride	Mn manganese	Se selenium	VitA vitamin A	Thia thiamine	
Ca calcium	K potassium	Su sulfur	Cu copper	I iodine	VitD vitamin D	Ribo riboflavin	

MIXED HAY

DIET DISCUSSION

Mixed hay (for the purpose of this discussion, half alfalfa and half timothy) results in a better caloric density for maintenance and light work, meaning the horse can eat more and be more satisfied. The calcium and phosphorus amounts and ratio are also good. As with all diets, the horse will need salt. Vitamin E will be deficient. Total crude protein is considered adequate by the NRC but lysine will be deficient. This combination is also borderline to deficient in many trace minerals and a supplement should be supplied. Mixed hay only is most appropriate for horses at maintenance or in light work.

* Insufficient to maintain weight at top expected intake.

DEFICIENCY ANALYSIS

■ (green) adequate ■ (yellow) marginal to inadequate ■ (red) inadequate ☐ (blank) level of this nutrient unknown

Activity	Nutriment	DE	CP	Ca	Ph	Mg	K	Na	C	Su	Fe	Mn	Cu	Z	Se	I	Co	VitA	VitD	VitE	Thia	Ribo	Lys

DE	digestible energy	Ph	phosphorus
CP	crude protein	Mg	magnesium
Ca	calcium	K	potassium
Na	sodium	Fe	iron
Cl	chloride	Mn	manganese
Su	sulfur	Cu	copper
Z	zinc	Co	cobalt
Se	selenium	VitA	vitamin A
I	iodine	VitD	vitamin D
VitE	vitamin E	Thia	thiamine
Ribo	riboflavin	Lys	lysine

SWEET FEEDS AND COMPLETE FEEDS

Many people feed their horse a mixture of grains with molasses—a sweet feed. All sweet feeds are definitely not created equal. One major difference that will be immediately obvious is the protein level. Sweet feeds are available ranging from 10% to 18% protein. What you may not realize is that along with the increased protein level you will be getting increased calories, as well as increased levels of the major and trace minerals. How much more of each of these the feed contains will depend on the protein source used.

Feeding a sweet feed mix is not the same as feeding plain oats or any other plain grains. Because sweet feeds vary in the amount of molasses they contain and the relative proportion of grains and other ingredients such as soybean meal, sweet feeds will vary widely in the levels of nutrients they contain. Your only source of information on sweet feeds will be the label on the bag and any further nutritional breakdown you can obtain from the manufacturer. You cannot assume that sweet feeds meet the NRC requirements for all nutrients—or even most of them. The list of ingredients may include many vitamins and minerals but does not tell you how much is actually in there. Statements such as "balanced," "perfect for," "designed for," etc., are basically meaningless. The only thing you can count on is what appears under the guaranteed analysis.

Guaranteed analysis will always include information on the % protein and % fiber. The lower the fiber, the more calories/energy the feed provides. Calcium and phosphorus content will also appear under guaranteed analysis. Almost all feeds also list vitamin A content and selenium content. Beyond this, the information you get can range from none to many or most of the other nutrients listed on our charts. The guaranteed analysis is just what it says—if you were to take the feed and have it analyzed it would contain the stated levels of those nutrients. The feed may also contain good levels of other nutrients but the manufacturer does not carefully check for those with each batch and supplement to keep them at a guaranteed level. In some cases, manufacturers do routinely supplement other nutrients based on what they know to be the average or usual levels in their feed ingredients but do not do close testing to make sure every single batch comes up to a certain level. The only way you can get an idea of what the average or usual level of these other nutrients is would be to contact the manufacturer and request a copy of a complete feed analysis. Some do not have this information. Others have it but won't release it. A few have it and are willing to give it to you. Unless you want to go to the trouble and expense of having a complete analysis done yourself, you should locate a manufacturer willing to share the information you need.

Complete feeds are not necessarily what the name implies. Even if they are advertised with statements such as "all your horse needs," you need to look closely at the label. Complete really only means that you do not have to feed any hay. The complete feed has enough fiber in it to keep the horse's digestive tract moving smoothly. Complete does NOT mean the feed measures up to NRC recommended levels for all important nutrients. As above, the only things you can count on are those listed under the guaranteed analysis. To learn more, you will have to contact the manufacturer—or ask your feed store to do it for you.

The complete feed will be similar to one of our grain/hay diets—either alfalfa, timothy (grass) or mixed hay and grain. If beet pulp was used instead of hay, magnesium will be lower than a diet based on alfalfa (although the ratio of calcium:magnesium will be better) and higher than a grass hay diet. Copper will be lower than with alfalfa or grass. Selenium will be borderline, zinc very low compared to hay diets. Manganese levels will be better than with alfalfa but not as good as with timothy hay based diets. Vitamin D levels are low. Of course, this can change if the manufacturer supplements one or more of these nutrients in the feed.

To show you what we mean, we took a look at several name brand commercial feeds. Where levels of certain key nutrients were not available, we had the feed analyzed by TPC Labs (The Pillsbury Company), St. Paul, MN. For the sweet feeds, we will show you how the complete ration works out (the feed plus alfalfa or timothy hay). For the complete feed, our charts show how they measure up to NRC recommendations when fed as is, without any supplements.

Stock Feed

You may have seen bags of feed with pictures of several different animals on the label and called simply Stock Feed or All Stock Feed. These are basically unsupplemented grain mixes, often with molasses added. Oats and corn form the base but other grains such as barley may also be used. Soybean meal is added to most. Their main attraction is the lower price compared to feeds sold specifically for horses. "Horse" feeds can be two, three or even more times expensive.

Can these be fed to horses? Yes, but there are several important differences you should know about. All stock feeds generally contain more protein (14 to 16%) than average horse feeds (10 to 12%). This is not a drawback. Magnesium and phosphorus levels tend to be slightly higher, also not a drawback—a plus if you are feeding alfalfa hay. The main problem is that the levels of copper, vitamin E and selenium in these feeds are too low for horses. Other trace mineral levels may also vary. It is fine to use these feeds, as long as you feed a complementary supplement. Your best course of action would be to ask the feed store representative to get you a complete analysis and suggest a supplement to match.

Do NOT use any feed that is labeled for use in cattle or poultry. These often contain special additives, antibiotics or other growth enhancers that can be toxic to horses. Feeds labeled as "All Stock" are free of these additives as a rule, but you should check with the feed store representative to make absolutely certain.

COMPLETE FEEDS

DIET DISCUSSION

So-called complete feeds vary widely in what they consider to be appropriate levels of nutrients. Complete feeds are often tailor made for certain categories of work. What works well for your racing, endurance, reining or event horse in heavy use will not work well in maintenance. This is because the energy level is too high in these feeds and you would have to cut way back on the amount fed to the horse that is not working to prevent his getting too fat. When you cut back on calories, you may also cut back on essential vitamins and minerals. In the following charts, blank boxes signify nutrients for which the manufacturer does not supply information, because either the level of nutrients cannot be guaranteed or the manufacturer has not checked nutrient levels. How to select a feed? Ask your feed store representative if a feed is designed for your class of horse and if it will meet the NRC recommendations for all vitamins, major and minor minerals. If they don't know, ask them to ask the manufacturer. See next page for details of discussion of feeds.

continued on next chart

DEFICIENCY ANALYSIS

Legend: ■ present in excessive amounts ■ adequate ■ marginal to inadequate ■ inadequate □ level of this nutrient unknown

Diet	Int	DM	Pro	Lys	DE	Ca	P	Mg	Na	Cl	K	Cu	Se	Z	I	Mn	VitA	VitD	VitE
Feed A	20																		
Feed A	39																		
Feed B	9																		
Feed B	17																		

Int Intake (lb/day)
DM Dry Matter (lb/day)
Pro Protein (lb/day)
Lys Lysine (oz/day)

DE Digestible Energy (Mcal/day)
Ca Calcium (g/day)
P Phosphorus (g/day)
Mg Magnesium (g/day)

Na Sodium (g/day)
Cl Chloride (g/day)
K Potassium (g/day)
Cu Copper (mg/day)

Se Selenium (mg/day)
Z Zinc (mg/day)
I Iodine (mg/day)
Mn Manganese (mg/day)

VitA Vitamin A (IU/day)
VitD Vitamin D (IU/day)
VitE Vitamin E (IU/day)

COMPLETE FEEDS—continued

DIET DISCUSSION

About the Feeds: Feed A is a complete feed designed for all classes of horses. However, you have to feed so much of it to a horse in heavy work that it is really not a very good choice.

Feed B is designed for competition horses. It will not work for horses at maintenance or low work levels because you feed so little of it that the horse will not get enough key minerals or protein.

Feed C is in between. It works best for performance horses because when fed at maintenance levels the amount you give is fairly small and horses may not be satisfied and could turn to vices such as wood chewing. However, it does supply all the needed nutrients even in the lesser amount.

Feed D is best for horses at maintenance or low work levels. They get to eat a sufficient volume to keep them happy. For the high performance horse, however, the amount you need to feed is probably more than they would willingly or comfortably take in in some cases.

Remember that with ALL diets, even those called "complete," your horse should have access to a plain salt block at all times.

DEFICIENCY ANALYSIS

Legend:
- ■ present in excessive amounts
- ■ adequate
- ■ marginal to inadequate
- ■ inadequate
- □ level of this nutrient unknown

Diet	Int	DM	Pro	Lys	DE	Ca	P	Mg	Na	Cl	K	Cu	Se	Z	I	Mn	VitA	VitD	VitE
Feed C	13																		
Feed C	25																		
Feed D	16																		
Feed D	33																		

Int   Intake (lb/day)
DM Dry Matter (lb/day)
Pro Protein (lb/day)
Lys Lysine (oz/day)

DE Digestible Energy (Mcal/day)
Ca   Calcium (g/day)
P   Phosphorus (g/day)
Mg Magnesium (g/day)

Na   Sodium (g/day)
Cl   Chloride (mg/day)
K   Potassium (g/day)
Cu   Copper (mg/day)

Se   Selenium (mg/day)
Z   Zinc (mg/day)
I   Iodine (mg/day)
Mn Manganese (mg/day)

VitA Vitamin A (IU/day)
VitD Vitamin D (IU/day)
VitE Vitamin E (IU/day)

SWEET FEEDS

DIET DISCUSSION

Here we take a look at the three top-of-the-line sweet feeds, each designed for performance horses, but with feeding directions for growing/pregnant horses and horses at maintenance or light work as well. In addition to the sweet feeds, all diets use hay (either timothy or alfalfa) at the rate of 1% of body weight (11 pounds a day) as the base. Again, there is a great variation between manufacturers concerning the level of specific nutrients. You will notice that toxic or close to toxic levels of vitamin A and vitamin D are present in each of these diets. These, and most other, commercial sweet feeds are highly fortified with these vitamins. This is done in part to accommodate the needs of growing or pregnant horses and in part because the actual vitamin content of the hay may be far below NRC averages. However, if you use supplemental vitamins, it is especially important to minimize or eliminate additional A and D when feeding a commercial feed, as toxicity problems could result. When you compare nutrient levels with these supplemented grains to those

continued on next chart

DEFICIENCY ANALYSIS

■ present in excessive amounts   ■ adequate   ■ marginal to inadequate   ■ inadequate   □ level of this nutrient unknown

Diet	Int	DM	Pro	Lys	DE	Ca	P	Mg	Na	Cl	K	Cu	Se	Z	I	Mn	VitA	VitD	VitE
Feed A	5																		
Feed A	14																		
Feed A	4																		
Feed A	14																		

Int Intake (lb/day)
DM Dry Matter (lb/day)
Pro Protein (lb/day)
Lys Lysine (oz/day)

DE Digestible Energy (Mcal/day)
Ca Calcium (g/day)
P   Phosphorus (g/day)
Mg Magnesium (g/day)

Na Sodium (g/day)
Cl Chloride (g/day)
K   Potassium (g/day)
Cu Copper (mg/day)

Se Selenium (mg/day)
Z   Zinc (mg/day)
I   Iodine (mg/day)
Mn Manganese (mg/day)

VitA Vitamin A (IU/day)
VitD Vitamin D (IU/day)
VitE Vitamin E (IU/day)

SWEET FEEDS—continued

DIET DISCUSSION

obtained when feeding plain oats, it is obvious they do a much better job of meeting most of the horse's needs. Notice also that the amount of grain recommended is actually quite small for horses at maintenance. With some horses, even this little bit of grain would result in weight gain. Notice also that the level of supplemental calcium in these grains is designed more for a grass hay based diet than alfalfa. While total calcium levels and calcium:phosphorus ratios are still within those described as tolerable by the NRC, the calcium is excessive and exceeds the ideal ratio.

DEFICIENCY ANALYSIS

Legend:
- ■ present in excessive amounts
- ■ adequate
- ■ marginal to inadequate
- ■ inadequate
- □ level of this nutrient unknown

Diet	Int	DM	Pro	Lys	DE	Ca	P	Mg	Na	Cl	K	Cu	Se	Z	I	Mn	VitA	VitD	VitE
Feed B	5																		
Feed B	16																		
Feed B	4																		
Feed B	15																		

Int Intake (lb/day)
DM Dry Matter (lb/day)
Pro Protein (lb/day)
Lys Lysine (oz/day)

DE Digestible Energy (Mcal/day)
Ca Calcium (g/day)
P Phosphorus (g/day)
Mg Magnesium (g/day)

Na Sodium (g/day)
Cl Chloride (g/day)
K Potassium (g/day)
Cu Copper (mg/day)

Se Selenium (mg/day)
Z Zinc (mg/day)
I Iodine (mg/day)
Mn Manganese (mg/day)

VitA Vitamin A (IU/day)
VitD Vitamin D (IU/day)
VitE Vitamin E (IU/day)

DEFICIENCY ANALYSIS

Legend:
- present in excessive amounts
- adequate
- marginal to inadequate
- inadequate
- level of this nutrient unknown

Diet	Int	DM	Pro	Lys	DE	Ca	P	Mg	Na	Cl	K	Cu	Se	Z	I	Mn	VitA	VitD	VitE
Feed C	5																		
Feed C	14																		
Feed C	4																		
Feed C	14																		

Abbreviations:

Int Intake (lb/day)
DM Dry Matter (lb/day)
Pro Protein (lb/day)
Lys Lysine (oz/day)

DE Digestible Energy (Mcal/day)
Ca Calcium (g/day)
P Phosphorus (g/day)
Mg Magnesium (g/day)

Na Sodium (g/day)
Cl Chloride (g/day)
K Potassium (g/day)
Cu Copper (mg/day)

Se Selenium (mg/day)
Z Zinc (mg/day)
I Iodine (mg/day)
Mn Manganese (mg/day)

VitA Vitamin A (IU/day)
VitD Vitamin D (IU/day)
VitE Vitamin E (IU/day)

12% PROTEIN ALL STOCK SWEET FEED

DIET DISCUSSION

Here we look at how a lower priced, lower protein feed compares to the higher priced brands. Again, hay is being fed at a rate of 1% of body weight—11 pounds per day. Our sample feed is a high quality 12% protein feed, compared to the 14 to 16% in the first sweet feeds we looked at. This is still a good quality feed but shows some problems that may be encountered using a supplemented grain, depending on a horse's level of work and the type and quality of hay used. Vitamin E and protein levels are borderline or too low to compensate for timothy hay when fed in maintenance amounts. Vitamin E level is too low to compensate for all classes and with either hay. Levels of zinc and manganese were not high enough to compensate for alfalfa hay when fed in maintenance amounts. Selenium may approach toxic levels when fed in the amount required by high performance horses (although most nutritionists would agree this 5 mg level is safe for high performance horses). These are not necessarily shortcomings in the feed or mistakes/omissions by the manufacturer. It is nearly impossible to formulate a supplemented feed for adult horses to meet all their requirements under all circumstances (different hays, different work levels).

DEFICIENCY ANALYSIS

Legend: ■ present in excessive amounts ■ adequate ▨ marginal to inadequate ▨ inadequate ☐ level of this nutrient unknown

Diet	Int	DM	Pro	Lys	DE	Ca	P	Mg	Na	Cl	K	Cu	Se	Z	I	Mn	VitA	VitD	VitE
Feed All	5																		
Feed All	17																		
Feed All	4																		
Feed All	16																		

Int Intake (lb/day)
DM Dry Matter (lb/day)
Pro Protein (lb/day)
Lys Lysine (oz/day)

DE Digestible Energy (Mcal/day)
Ca Calcium (g/day)
P Phosphorus (g/day)
Mg Magnesium (g/day)

Na Sodium (g/day)
Cl Chloride (g/day)
K Potassium (g/day)
Cu Copper (mg/day)

Se Selenium (mg/day)
Z Zinc (mg/day)
I Iodine (mg/day)
Mn Manganese (mg/day)

VitA Vitamin A (IU/day)
VitD Vitamin D (IU/day)
VitE Vitamin E (IU/day)

FACTORS AFFECTING MINERAL LEVELS IN FEEDS

PLANT FACTORS

Plants will tend to be highest in the minerals they most require to complete their life cycles. Since different species of plants have different requirements, this leads to many of the differences we see in mineral content of different types of hay or grain. Alfalfa has three times as much calcium as timothy hay, but timothy has double the amount of manganese. Within the plant itself, various portions—leaves, stems, roots and grains/seeds—also have different mineral requirements and will vary in their levels. For example, oats contain only 0.14% magnesium, but hay made from the plant itself contains 0.26% magnesium. If the oat hay is allowed to completely age, dry and weather until it becomes straw, the magnesium content again drops, down to 0.16% on the average.

SOIL FACTORS

The mineral composition of the soils varies geographically. The United States Plant, Soil and Nutrition Laboratory has mapped out several mineral-deficient areas, and this information has proven very useful to farmers in growing their crops and livestock.

Cobalt, essential to the production of vitamin B-12, is deficient in soils of New England and along the Atlantic and Gulf coasts. Copper, essential for formation of red blood cells, normal reproduction, pigmentation of the coat and in important antioxidant systems, is deficient in some soils and in others its absorption is blocked by high levels of the mineral molybdenum (Florida and some western states). Soils of the Pacific Northwest, states along the Great Lakes, areas of Canada, the eastern seaboard and most of Florida are deficient in the trace mineral selenium. This has resulted in weakness

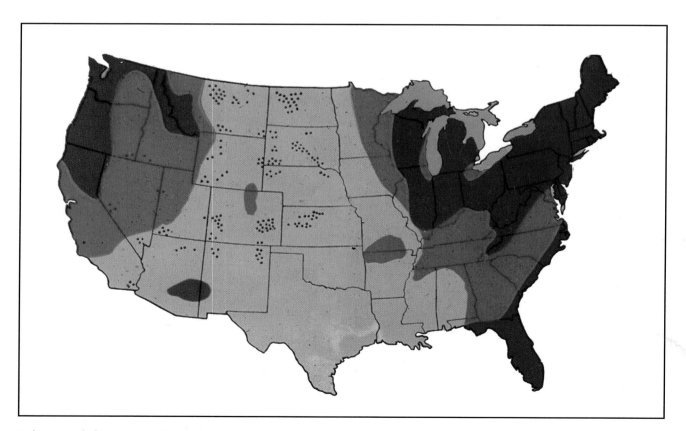

Selenium deficiency in the soil, and therefore in grains and hays, is a widespread problem in the U.S.

Copper map

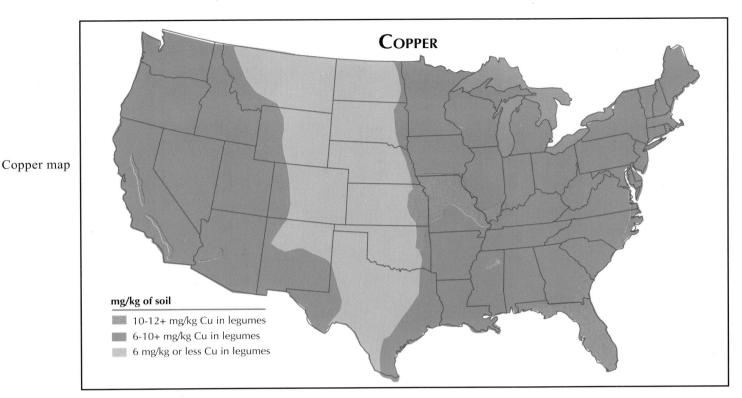

COPPER

mg/kg of soil

- 10-12+ mg/kg Cu in legumes
- 6-10+ mg/kg Cu in legumes
- 6 mg/kg or less Cu in legumes

Copper deficiency is a widespread problem in equine diets and has been linked to serious joint problems (osteochondrosis dessicans) in young foals. As this map shows, only relatively small areas of the U.S. can grow legumes (e.g., alfalfa) that contain even the minimum required for adult horses (10 to 13 mg/kg).

and deaths in calves and foals caused by white muscle disease, decreased fertility and may contribute to tying-up. Chickens are also extremely sensitive to selenium deficiency. Work is currently underway to map soils which are deficient in zinc and nickel. Nickel has only recently been recognized as essential to the normal growth and health of plants and animals.

Fertilizers will also influence the mineral levels in the soils. Plants will be higher in the minerals contained in the fertilizer that are beneficial to their life cycles. In most cases this will be potassium or phosphorus, depending on the composition of the fertilizer.

ACID RAIN

Most of us have heard at least something about acid rain—how it destroys the paint jobs on our cars and eats away at statues, monuments and historical sites. For the most part, however, acid rain does not have much of an impact on our daily lives.

A less publicized aspect of acid rain is the dam-

age is does to soils, plants and water supplies in affected areas. Acid rain is more than just a doomsday warning from radical environmentalists. The United States government has taken acid rain seriously enough to enact laws forcing industry to decrease automobile exhaust emissions responsible for acid rain and to set up monitoring programs across the country. Acid rain results when emissions (largely from the burning of fossil fuels) from industry, sulfur and nitrogen compounds, react with the oxygen in the air. Soils with high buffering capacity (ability to neutralize the acidity), such as those in the many Midwest states, can handle the extra acid load without adverse effects. Soils with poor buffering capacity, such as those east of the Mississippi and especially in states along the eastern seaboard, will not be able to neutralize the acidity well, and changes will occur in the minerals of the soil.

Acid rain that is not neutralized will dissolve beneficial nutrients such as calcium, magnesium and potassium, causing them to be washed away. At the same time, toxic minerals such as aluminum, which are in a stable, bound form at higher

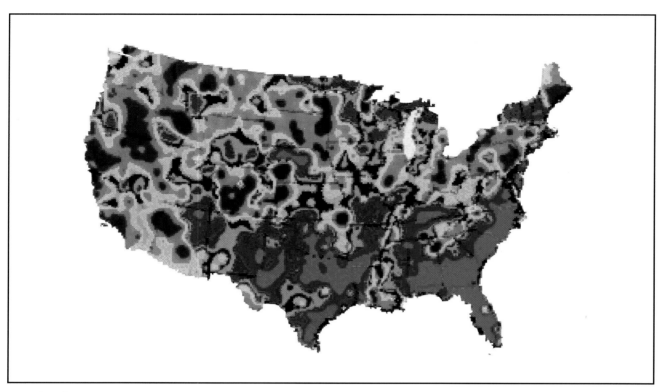

Zinc levels vary over an extremely wide range in U.S. soils, but how much is available to the plant depends on a number of factors including temperature, pH (acidity) and level of phosphorus in the soil.

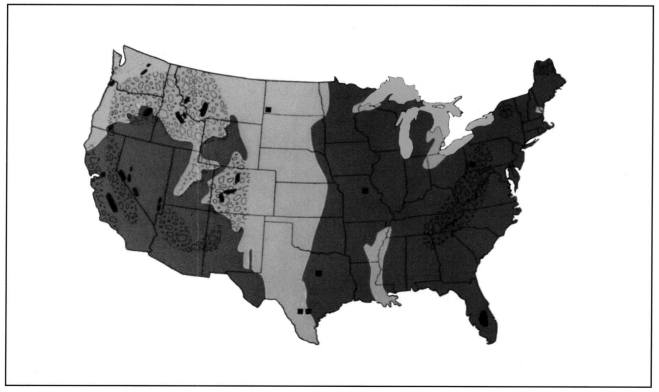

High molybdenum levels are known to interfere with absorption of copper by many species of livestock. Although horses can tolerate much higher levels than cattle without developing problems, the alarming rise in soil levels caused by industrial pollution can result in problems for horses as well.

Equine Supplements & Nutraceuticals

soil pH, become liberated and will substitute for some of the other nutrients that are washed away. The higher concentration of aluminum also appears in the ground water of affected areas. Similar changes occur to surface waters, such as lakes. The end result is less availability of beneficial minerals as they become washed away and higher availability of toxic minerals such as aluminum.

Farmers can offset some of these changes by treating their soils with limestone (a buffering agent) and fertilizers which are high in the beneficial minerals. Crops grown on treated soils are less likely to be affected. However, changes may still occur in grazing land and in the water supply. There is evidence from hair mineral analysis (see also Chapter 7, Consumer's Guide to Supplements) that aluminum toxicity is a growing problem. Toxic levels of aluminum are appearing with alarming frequency in livestock and horse hair samples reported from Uckele Animal Health and Nutrition Laboratories in Blissfield, Michigan. In some cases this can be traced to contamination of the premises (bases used in making livestock feed), but in others, most notably in horses, it cannot. Horses are believed to be very sensitive to aluminum in the environment and are an indicator species—accumulating high levels of aluminum very easily.

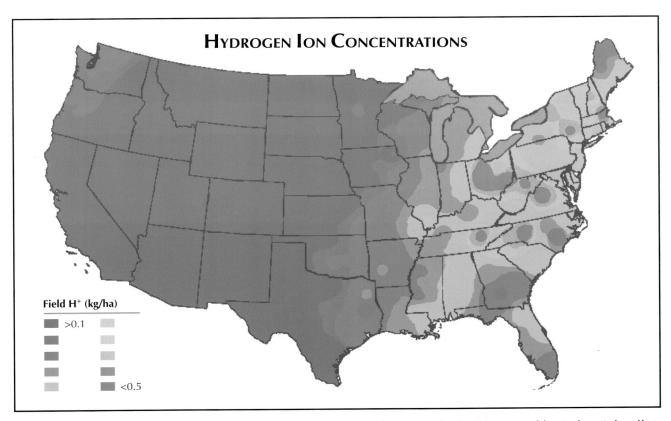

HYDROGEN ION CONCENTRATIONS

Field H⁺ (kg/ha)

>0.1

<0.5

Rising acidity (low pH levels) in the rain, referred to commonly as "acid rain," is caused by industrial pollutants in the air and is robbing our soil of many essential nutrients.

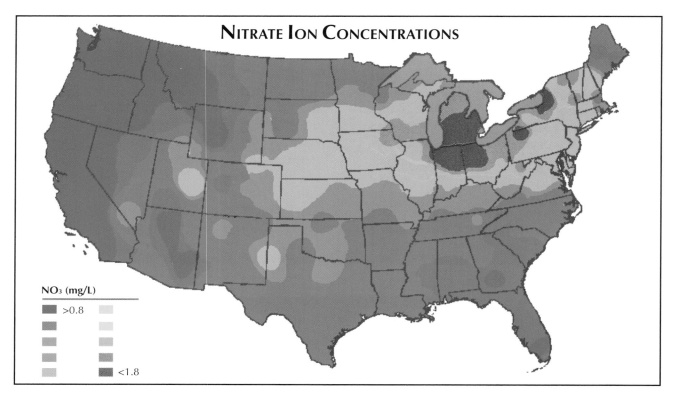

NITRATE ION CONCENTRATIONS

NO₃ (mg/L)

>0.8

<1.8

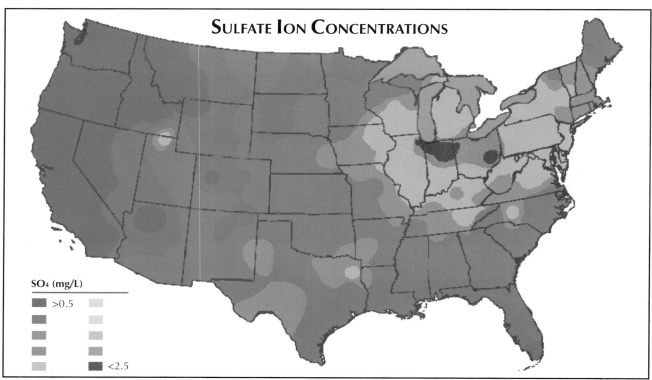

SULFATE ION CONCENTRATIONS

SO₄ (mg/L)

>0.5

<2.5

Nitrate and sulfate ions are two of the pollutants found in acid rain which are responsible for detrimental changes to the mineral levels in our soils.

CHAPTER 2

MARES AND GROWING HORSES

The nutrients in a mare's milk are almost 100 percent absorbed by her foal.

ABOUT THIS CHAPTER

The pregnant mare, weanling, yearling and growing horse through the second year of life have special nutritional needs in terms of energy, protein, vitamin and mineral content of the diet. The requirements for growth and the building of strong, healthy tissues are very different from those for any other phase of life, even heavy work. Failure to meet these needs can be nothing short of disastrous. If you are involved in the care of a pregnant mare or young horse, you will need to be committed to learning and carefully following their dietary needs.

The following table by L. A. Lawrence, Extension animal scientist, Horses, Virginia Tech, will give you some idea of how quickly and dramatically the nutritional needs change with age and even rate of growth. In addition, the functions and requirements for the trace minerals and vitamins will be different and even more important than in an older animal, with consequences (often damage) existing for the life of the horse.

CONSEQUENCES OF INADEQUATE NUTRITION

We would all expect that a fetus or young horse that is not given enough to eat will not grow well. This is indeed true but is only the tip of the iceberg. Foals that are fed too much are also prone to

FAT—All horses can survive on very low fat levels in their diet. However, a certain minimal amount of the essential fatty acids (see Chapter 4, A to Z Other, Essential Fatty Acids) is needed to maintain general health and proper production of hormones. No one pays any particular attention to fat content or sources in the diet of pregnant, lactating or growing horses—unless it is being used as an additional source of calories. Many of the high quality foods (grains, protein sources, pasture) commonly used in diets for these groups probably suffice to provide all the needed essential fatty acids. For added insurance you can provide small amounts of unprocessed vegetable oils.

VITAMINS AND MINERALS

Calcium and phosphorus: The table shows the wide variation in requirements for these two nutrients in mares and horses of various ages. Needs are linked to growth of the fetus, production of milk and growth of the skeleton in the growing horse. Failure to provide adequate calcium to milking mares can cause a serious muscular abnormality, milk tetany, that leads to profound weakness, irregularities of heart beat and potentially even death of the mare. However, providing too much too soon (i.e., before she foals) can cause the hormone systems that mobilize calcium to meet milk demands to become too sluggish, again resulting in the same problem of milk tetany.

Calcium and phosphorus are also the major minerals involved in normal formation of strong, healthy bones and joints. The ration must be balanced not only in terms of the total amount of each that is fed but to ensure that the ratio between them remains close to or within the ideal range of about 1.5:1, calcium:phosphorus.

Magnesium: Magnesium is also found largely in bone but plays important functions in the muscle and nervous system. Too little magnesium can contribute to the likelihood of milk tetany. Inadequate magnesium at any age may cause problems of nervousness and muscular weakness or cramping.

Trace Minerals: Copper, zinc and manganese are absolutely critical to the formation of normal bones and especially joints. Many nutritionists recommend feeding levels of copper and zinc that significantly exceed the NRC levels—up to 40 ppm for copper and 120 ppm for zinc. A 1998 paper from New Zealand confirms levels of 0.5 mg/kg of body weight a day in the diets of pregnant horses helps prevent OCD. Diets high in iron (over 40 ppm) will block absorption of these key minerals, leading to increased requirements. These same minerals are essential to the development of a strong immune system and for health of all connective tissues throughout the body.

Selenium is required for a normal nervous system, immune system and for muscular function. Deficiency results in a disease in foals called white muscle disease in which the muscles are pale or even calcified and fatality is common.

Iodine is needed for normal thyroid function and therefore normal pregnancy and growth. Feed in usual required amounts.

Vitamins: Vitamin A is required for normal formation of the eye, skin, hair, hooves and bone. This is one of the most frequently supplemented vitamins and will be adequate in any commercial diet, alfalfa hay or on pasture. Mares or growing animals kept stabled on grass hay only may need additional vitamin A.

Vitamin D is needed for normal development of bone. This is the sunshine vitamin and does not need to be supplemented if mares and young horses get exposure to sunlight—their bodies will make it.

Vitamin C—again, we know very little about vitamin C in the horse or how much benefit it could be to pregnant and growing animals. However, we do know that stabling and eating of hay only (no pasture) diets results in drops in vitamin C to virtually nondetectable levels in blood. We also know that vitamin C is essential to the health of the immune system and of all connective tissues—including joint cartilage, tendons and ligaments. Every time it has been studied in other species, vitamin C has resulted in improved strength in those tissues. Personally, I take the safe route and supplement it—4.5 grams per day for mares and growing animals.

B Vitamins: There is little information on B vitamin needs for pregnancy and growth/development. B vitamins are involved in many body processes. Most important to the pregnant mare and growing animal is the role of B vitamins in normal development of the brain, spinal cord and nerves and in metabolism of protein and carbohydrates. One study showed significantly better weight gain in young ponies fed a thiamine supplemented diet.

No problems have been recognized in mares and growing animals fed normal diets with no B vitamin supplementation. On the other hand, we don't know if they would have benefitted either. B vitamin supplementation is optional (as with vitamin C). Supplementation up to the NRC recommended daily intake—i.e., supplementing with amounts equivalent to what the horse should take in on a daily basis, much like a human taking a multivitamin—is safe. (See Chapter 3, Nutrition A to Z)

SAMPLE DIETS

The following charts are set up in the same way as the feeding charts in Chapter 1, Basic Nutrition. All diets were balanced first for energy. That is, the energy/calorie requirements were determined and the amount of feed appropriate to maintain body weight and support growth or lactation is determined. The other nutrient levels refer to a total intake sufficient to meet the energy needs of growth, pregnancy, etc. Specific categories of horse include:

- Mares—the weight of the horse is 1,100 lbs (454.54 kg). For growing animals, expected mature weight is 500 kg.
- Weaning, 4 months—weight 385 lbs or 175 kg, average daily gain 1.87 lbs or 0.85 kg/day.
- Weanling, 6 months, moderate growth rate—weight 473 lbs or 215 kg, average daily gain 1.65 lbs or 0.75 kg/day.
- Yearling, moderate growth rate—weight 715 lbs or 325 kg, average daily gain 1.1 lbs or 0.5 kg/day.
- Long Yearling, 18 months, moderate growth rate—weight 880 lbs or 400 kg, average daily gain 1.77 lbs or 0.35 kg/day.
- 2 year old—990 lbs or 450 kg, average daily gain 0.44 lbs or 0.2 kg/day.
- 2 year old in training—same as above.
- Pregnancy last trimester
- Early lactation
- Late lactation

NOTE: In Basic Nutrition, diets were calculated with proportions of hay:grain on a weight basis. That is, 50:50 oats and alfalfa meant equal weights of oats and alfalfa were fed. In these diets, the percent contribution of the grains and hays was on a calorie basis. That is, a diet of 50% hay and 50% grain would have 50% of the calories coming from hay and 50% from the grain. Since grain is a more dense source of calories, the amount of grain being fed would weigh less than the amount of hay.

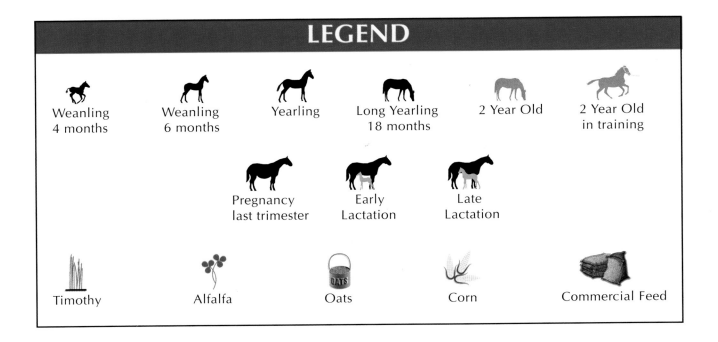

LEGEND

Weanling 4 months | Weanling 6 months | Yearling | Long Yearling 18 months | 2 Year Old | 2 Year Old in training

Pregnancy last trimester | Early Lactation | Late Lactation

Timothy | Alfalfa | Oats | Corn | Commercial Feed

WEANLING - 4 MONTHS

DIET DISCUSSION

Rapidly growing horses have very high energy and mineral demands. Errors at this stage of growth can result in permanent damage to bones and joints. Because you are feeding relatively small amounts compared to what adults will eat, and a higher proportion of grain, errors in the amount fed can have a great impact on the nutrient levels in the diet, shifting key minerals, protein and lysine from adequate to deficient. Investing in a properly supplemented grain mix, following directions closely and making sure to actually weigh the grain, not estimate, eliminates the guesswork and results in a properly balanced diet.

NOTE: Although more research data needs to be gathered to confirm the benefit, studies suggest that additional copper in the diet of growing horses may help prevent bone and joint disease. Because of this, our recommendations differ from the NRC's on copper and this book uses a level three times that of the NRC. As a result, zinc needs also changed to maintain zinc:copper at a proper level.

This diet provides 70 percent of the calories from grain and 30 percent from hay, as suggested in the NRC guidelines.

DEFICIENCY ANALYSIS

■ (green) adequate   ▨ (yellow) marginal to inadequate   ■ (red) inadequate   □ (blank) level of this nutrient unknown

Diet	Nutriment	DE	CP	Ca	Ph	Mg	K	Na	Cl	Su	Fe	Mn	Cu	Z	Se	I	Co	VitA	VitD	VitE	Thia	Ribo	Lys
1																							
2																							
3																							
4																							
5																							
6																							

DE digestible energy   Ph phosphorus   Na sodium   Fe iron   Z zinc   Co cobalt   VitE vitamin E   Lys lysine
CP crude protein   Mg magnesium   Cl chloride   Mn manganese   Se selenium   VitA vitamin A   Thia thiamine
Ca calcium   K potassium   Su sulfur   Cu copper   I iodine   VitD vitamin D   Ribo riboflavin

WEANLING - 6 MONTHS

DIET DISCUSSION

As the young horse grows, his digestive tract becomes more like an adult's. Therefore, the percentages of grain and hay being fed should gradually change. Nutritional requirements for the 6 month old weanling are still quite high and impossible to meet with a plain hay and grain diet. Your wisest choice remains a properly supplemented feed and alfalfa hay. Again, these diets recommend

higher than NRC's currently recommended level for copper and a proportional increase in zinc as well.

This diet provides 60 percent of calories from grain and 40 percent from hay, as suggested in the NRC guidelines.

DEFICIENCY ANALYSIS

■ (green) adequate   ▨ (yellow) marginal to inadequate   ■ (red) inadequate   ☐ (blank) level of this nutrient unknown

Nutriment

Diet	DE	CP	Ca	Ph	Mg	K	Na	Cl	Su	Fe	Mn	Cu	Z	Se	I	Co	VitA	VitD	VitE	Thia	Ribo	Lys
1																						
2																						
3																						
4																						
5																						
6																						

DE digestible energy
CP crude protein
Ca calcium
Ph phosphorus
Mg magnesium
K   potassium
Na sodium
Cl chloride
Su sulfur
Fe iron
Mn manganese
Cu copper
Z   zinc
Se selenium
I   iodine
Co cobalt
VitA vitamin A
VitD vitamin D
VitE vitamin E
Thia thiamine
Ribo riboflavin
Lys lysine

YEARLING

DIET DISCUSSION

By the age of a year, a horse eats more adult proportions of hay and grain. Although he weighs quite a bit more, his growth is slowing down so that the calorie requirements at this age are not much different than those of the 6 month old weanling. As the charts clearly show, your best feeding option is still a properly supplemented grain mix and alfalfa hay. Again, our charts reflect copper

requirements 3 times NRC current recommendations and increased zinc to balance the extra copper.

This diet provides 60 percent of the calories from grain and 40 percent from hay, as suggested in the NRC guidelines.

DEFICIENCY ANALYSIS

■ (green) adequate   ■ (yellow) marginal to inadequate   ■ (red) inadequate   □ (blank) level of this nutrient unknown

Diet	Nutriment	DE	CP	Ca	Ph	Mg	K	Na	Cl	Su	Fe	Mn	Cu	Z	Se	I	Co	VitA	VitD	VitE	Thia	Ribo	Lys
1																							
2																							
3																							
4																							
5																							
6																							

DE digestible energy   Ph phosphorus   Na sodium   Fe iron   Z zinc   Co cobalt   VitE vitamin E   Lys lysine
CP crude protein   Mg magnesium   Cl chloride   Mn manganese   Se selenium   VitA vitamin A   Thia thiamine
Ca calcium   K potassium   Su sulfur   Cu copper   I iodine   VitD vitamin D   Ribo riboflavin

LONG YEARLING (18 MONTHS)

DIET DISCUSSION

With changing proportions of hay and grain, each take on a different importance in the diet. Changing from oats to oats and corn is enough to shift nutrients into the red zone. Stay with the fully supplemented grain mix and alfalfa for best results. Copper and zinc recommendations remain above NRC.

Diet 2 is a mix of timothy and 50:50 oats and corn. Diet 3 is a mix of alfalfa and 50:50 oats and corn.

This diet provides 55 percent of the calories from hay and 45 percent from grains, as suggested in the NRC guidelines.

DEFICIENCY ANALYSIS

■ (green) adequate   ▨ (yellow) marginal to inadequate   ■ (red) inadequate   □ (blank) level of this nutrient unknown

Diet	DE	CP	Ca	Ph	Mg	K	Na	Cl	Su	Fe	Mn	Cu	Z	Se	I	Co	VitA	VitD	VitE	Thia	Ribo	Lys
1																						
2																						
3																						
4																						
5																						
6																						

DE digestible energy   Ph phosphorus   Na sodium   Fe iron   Z zinc   Co cobalt   VitE vitamin E   Lys lysine
CP crude protein   Mg magnesium   Cl chloride   Mn manganese   Se selenium   VitA vitamin A   Thia thiamine
Ca calcium   K potassium   Su sulfur   Cu copper   I iodine   VitD vitamin D   Ribo riboflavin

Nutriment

2 YEAR OLD

DIET DISCUSSION

A 2 year old, not in training, is recommended to receive 35% of the diet as grain, 65% as hay. It is impossible to correctly feed a 2 year old at these ratios without using a supplement, even using supplemented grains. This is because the percentage contributed by the grain has dropped to a lower level. For these calculations, NRC recommended levels of copper were used (but many diets could not even measure up to those).

Diet 2 is 50:50 timothy and oats/corn.
Diet 3 is 50:50 alfalfa and oats/corn.

This diet provides 35 percent of the calories from grain and 65 percent from hay, as suggested in the NRC guidelines.

DEFICIENCY ANALYSIS

■ (green) adequate ■ (yellow) marginal to inadequate ■ (red) inadequate ☐ (blank) level of this nutrient unknown

Diet	DE	CP	Ca	Ph	Mg	K	Na	Cl	Su	Fe	Mn	Cu	Z	Se	I	Co	VitA	VitD	VitE	Thia	Ribo	Lys
1																						
2																						
3																						
4																						
5																						
6																						

Nutriment

DE digestible energy
CP crude protein
Ca calcium
Ph phosphorus
Mg magnesium
K potassium
Na sodium
Cl chloride
Su sulfur
Fe iron
Mn manganese
Cu copper
Z zinc
Se selenium
I iodine
Co cobalt
VitA vitamin A
VitD vitamin D
VitE vitamin E
Thia thiamine
Ribo riboflavin
Lys lysine

2 YEAR OLD IN TRAINING

DIET DISCUSSION

The NRC recommends that 2 year olds in training receive 1.39 times as much feed as those not in training and that the proportions of hay and grain be changed to 50:50. This improves the picture when using supplemented grains, but you will still need an additional vitamin supplement.

Diet 2 is a mix of timothy and 50:50 oats and corn.
Diet 4 is a mix of alfalfa and 50:50 oats and corn.

DEFICIENCY ANALYSIS

■ (green) adequate   ■ (yellow) marginal to inadequate   ■ (red) inadequate   □ (blank) level of this nutrient unknown

Diet	DE	CP	Ca	Ph	Mg	K	Na	Cl	Su	Fe	Mn	Cu	Z	Se	I	Co	VitA	VitD	VitE	Thia	Ribo	Lys
1																						
2																						
3																						
4																						
5																						
6																						

DE digestible energy   Ph phosphorus   Na sodium   Fe iron   Z zinc   Co cobalt   VitE vitamin E   Lys lysine
CP crude protein   Mg magnesium   Cl chloride   Mn manganese   Se selenium   VitA vitamin A   Thia thiamine
Ca calcium   K potassium   Su sulfur   Cu copper   I iodine   VitD vitamin D   Ribo riboflavin

DIET DISCUSSION

Weight 1,100 lbs, 30% grain and 70% hay

Adequate amounts and proper ratios of calcium, phosphorus, protein, lysine and all trace minerals are critical for the heavily pregnant mare. A very high quality diet is needed since abdominal pressure may prevent the mare from eating large amounts. Alfalfa hay (with adequate phosphorus added to the ration) is preferred since she will not have to eat as much. High quality, supplemented, 14% grain mixes avoids many common deficiencies found with plain grains. As is often the case, feed companies formulate their 14% protein feed to meet the needs of performance horses and some, but not all, stages of growth and pregnancy. It is impossible to make one feed do all these specialized jobs. They, and others, also manufacture a high protein vitamin and mineral supplement to add to the ration to meet different requirements of various stages of pregnancy and growth.

NOTE: NRC guidelines were used for all elements except copper. A 1998 paper from New Zealand confirms additional copper is effective in decreasing the incidence of OCD in foals. The copper requirement used in these diets was 3 times the NRC current recommendation. Because of this, zinc requirements also had to be changed to keep the zinc:copper ratio in the desired range.

DEFICIENCY ANALYSIS

■ (green) adequate   ■ (yellow) marginal to inadequate   ■ (red) inadequate   □ (blank) level of this nutrient unknown

Diet	DE	CP	Ca	Ph	Mg	K	Na	Cl	Su	Fe	Mn	Cu	Z	Se	I	Co	VitA	VitD	VitE	Thia	Ribo	Lys
1																						
2																						
3																						
4																						
5																						
6																						

DE digestible energy   Ph phosphorus   Na sodium   Fe iron   Z zinc   Co cobalt   VitE vitamin E   Lys lysine
CP crude protein   Mg magnesium   Cl chloride   Mn manganese   Se selenium   VitA vitamin A   Thia thiamine
Ca calcium   K potassium   Su sulfur   Cu copper   I iodine   VitD vitamin D   Ribo riboflavin

DIET DISCUSSION

The calorie, protein, lysine, vitamin and mineral needs of a mare in early lactation are greater than for any other category. In fact, a mare that enters this stage without a good supply of body fat is likely to lose considerable weight. The highest quality hay and grain are an absolute necessity for both health of the mare and the quality of her milk—the ONLY food the rapidly growing foal will be getting. Feeding alfalfa and a properly supplemented grain can eliminate your need for additional supplements.

A diet of alfalfa or timothy hay only requires an intake of 30 lbs a day.

This diet provides 60 percent of its calories from hay and 40 percent from grain, as suggested in the NRC guidelines.

Weight: 1,100 pounds

DEFICIENCY ANALYSIS

■ (green) adequate   ■ (yellow) marginal to inadequate   ■ (red) inadequate   ☐ (blank) level of this nutrient unknown

Diet	Nutriment DE	CP	Ca	Ph	Mg	K	Na	Cl	Su	Fe	Mn	Cu	Z	Se	I	Co	VitA	VitD	VitE	Thia	Ribo	Lys
1																						
2																						
3																						
4																						
5																						
6																						

DE digestible energy   Ph phosphorus   Na sodium   Fe iron   Z zinc   Co cobalt   VitE vitamin E   Lys lysine
CP crude protein   Mg magnesium   Cl chloride   Mn manganese   Se selenium   VitA vitamin A   Thia thiamine
Ca calcium   K potassium   Su sulfur   Cu copper   I iodine   VitD vitamin D   Ribo riboflavin

LATE LACTATION

DIET DISCUSSION

As lactation slows, it becomes theoretically possible to meet the mare's energy needs with hay alone although the amount required is quite large (just over 20 lbs of alfalfa, 28 lbs of timothy). Better to go with a high protein, high lysine supplemented grain mix. That same mix will also work for creep feeding and for weaning.

The diet provides 35 percent of its calories from grain and 65 percent from hay, as suggested in NRC guidelines.

DEFICIENCY ANALYSIS

■ (green) adequate   ■ (yellow) marginal to inadequate   ■ (red) inadequate   ☐ (blank) level of this nutrient unknown

Nutriment

Diet	DE	CP	Ca	Ph	Mg	K	Na	Cl	Su	Fe	Mn	Cu	Z	Se	I	Co	VitA	VitD	VitE	Thia	Ribo	Lys
1																						
2																						
3																						
4																						
5																						
6																						

DE digestible energy
CP crude protein
Ca calcium
Ph phosphorus
Mg magnesium
K potassium
Na sodium
Cl chloride
Su sulfur
Fe iron
Mn manganese
Cu copper
Z zinc
Se selenium
I iodine
Co cobalt
VitA vitamin A
VitD vitamin D
VitE vitamin E
Thia thiamine
Ribo riboflavin
Lys lysine

CHAPTER 3

NUTRITION A TO Z

The well nourished horse is always interested in her surroundings.

ABOUT THIS CHAPTER

This chapter discusses the major vitamins and minerals recognized by the National Research Council as being required in the diets of horses. It is not a guide to supplementation per se, although it will give you an idea of the total daily requirement for these nutrients based upon the horse's diet and level of activity.

This figure estimates total daily need for the vitamin or mineral from all sources—hay, grain, pasture, even water, as well as supplements. How much, if any, should be added as a supplement will depend on the basic diet and use of the horse.

This chapter will describe each nutrient, sources in the diet, conditions which result in increased need or decrease the amount available to the horse, possibility of overdosage/toxicity/adverse reactions, other nutrients which play complementary roles and possible symptoms you might see if a deficiency or less than optimal level of intake exists.

After reading this section you may be interested in several specific vitamins or minerals. To determine if your horse's basic diet may be inadequate, consult Chapter 1, Basic Nutrition, which looks in detail at specific diet types and which vitamins and minerals are borderline, adequate or low by NRC standards. You should also check out Chapter 6, Relieving Health Problems Through Nutrition, for guidance regarding different problems that may be caused or worsened by inadequate vitamin/mineral intake.

The recommended daily intake amounts in this chapter are based on information from the NRC Nutrient Requirements for Horses as well as extrapolation from information available from other species, particularly humans, since most of the information regarding nutrition and performance is from human studies. Obviously people and horses eat very different diets, which means the amount of vitamins and minerals they consume in average diets will be quite different. They also have vastly different digestive tracts, which will influence how much of what they take in can actually be digested and/or absorbed. There may even be differences in requirements for specific nutrients that are species related—with horses needing a high percentage of this in their diet and humans more of that. However, the basic ways in which all species respond to specific nutrients—vitamin C for example—are remarkably similar. Many vitamins and minerals do the same things in people as they do in horses. This forms the backbone of comparative medicine, a study of the similarities and differences in normal physiology and disease states in various species.

Because tissues respond in the same way to the basic building blocks, vitamins and minerals in foods, nutritionists, physiologists and supplement manufacturers will often fill in the blanks by using information from one species to estimate daily requirements of another. When converting human requirements to equine, a factor of 6 is used—i.e., horses will need 6 times as much under similar circumstances. This factor is based on metabolic body weight—a figure that takes into account both body weight and surface area of the body to determine, in essence, how much of the weight represents metabolically active tissues. Converting a human requirement to equine simply on the basis of body weight alone would lead to overestimating.

The next chapter, A to Z Other, will cover nutrients with no established NRC level of intake.

LEGEND

Use Icons	Probability of Deficiency					Severity of Deficiency & Risk of Toxicity	Diet Icons
Maintenance	None	Little	Moderate	High	Heavy	None	Pasture
Light	None	Little	Moderate	High	Heavy	Little	Timothy
Moderate	None	Little	Moderate	High	Heavy	Moderate	Alfalfa
High	None	Little	Moderate	High	Heavy	Serious	Oats
Heavy	None	Little	Moderate	High	Heavy		

BIOTIN

B VITAMIN

Probability of Deficiency Per Use	RDA Per Use	Severity of Deficiencies	Toxicity Risk	Complementary Element	Symptoms of Deficiency
	 0			Magnesium	• Brittle, shelly outer horn of hoof
	 0				• Dry skin
	 0				• Poor hair coat
	 0				• Possible elevated blood sugar
	 0				

Supplementation Recommendation

USE / DIET					
(hay)	0-20mg	0-20mg	0-20mg	0-20mg	0-20mg
(hay + bucket)	0	0	0	0	0
(clover)	0-20mg	0-20mg	0-20mg	0-20mg	0-20mg
(clover + bucket)	0	0	0	0	0
(hay + clover)	0-20mg	0-20mg	0-20mg	0-20mg	0-20mg
(hay + clover + bucket)	0	0	0	0	0

Sources

The horse receives B vitamins in his diet from such things as grains, brans and yeast. In addition, the abundant microorganism population in the horse's intestinal tract is capable of manufacturing B vitamins and it is believed the horse can absorb a certain amount from this source as well. Hays supply essentially no biotin although it is believed that absorption of biotin manufactured in the gut will supply an adequate amount.

Functions

The B vitamins play essential roles in virtually every organ and cell of the horse's body. Biotin is involved in glucose metabolism, growth, utilization of niacin, another B vitamin, and maintenance of all rapidly dividing tissues. In horses, this sulfur containing vitamin is believed to be especially important in the maintenance of healthy hooves.

Supplementation

The B vitamins belong to the group called water soluble vitamins. What this means is that they float freely through the fluids of the body and are not stored in any body tissues. Once these vitamins are absorbed into the blood, they will circulate and be taken in by cells that need them or eliminated in the urine. Because of this rapid elimination, the horse must take in needed B vitamins on a constant, daily basis. Small amounts may be stored in the liver. Biotin is synthesized in the intestines, but levels do not begin to rise until you reach the large intestine. Since absorption is likely to be best from the small intestine, the contribution to total biotin intake from this source is questionable.

Deficiencies

Full blown, life-threatening deficiency symptoms for B vitamins are not likely to develop in horses on normal diets. Even if a horse is not eating, the organisms in the intestinal tract are capable of manufacturing these vitamins and some will be absorbed by the horse. However, the horse can develop symptoms of inadequate amounts of B vitamins. Specifically, biotin has been shown to result in improved hoof quality in a certain percentage of horses with shelly, brittle feet. Biotin is not the only nutrient associated with poor hoof quality, however. Only an estimated 2% of horses with hoof problems have an uncomplicated biotin deficiency. Zinc is also important to healthy hooves, as are the amino acids methionine and lysine and all vitamins and minerals required for healthy connective tissues that attach the outer hoof wall to the living tissue underneath. For best results with problem feet, all nutrients should be supplemented. (See reference to hoof problems in Chapter 6, Relieving Health Problems Through Nutrition.)

Toxicity

Toxicity with B vitamins administered in the feed is virtually nonexistent since the kidneys will eliminate any excess very efficiently. There is no known biotin toxicity in any species.

Interactions

Many of the B vitamins work in conjunction so it is usually advised that if you supplement with one B vitamin you should also supplement the others.

Indications

In general, B vitamin supplementation is indicated in any horse that is not eating normally (or not eating at all), in horses that are heavily stressed for any reason (age, surgery, injury, infection, shipping, etc.) and in horses that are being heavily exercised. Horses with poor quality hooves may benefit from biotin supplementation at 10 to 20 mg/day, higher doses being indicated for horses receiving little to no grain. However, as discussed above, uncomplicated biotin deficiency as a cause of poor quality feet accounts for a relatively small percentage of all horses with hoof problems.

CALCIUM

MAJOR MINERAL

Probability of Deficiency Per Use	RDA Per Use	Severity of Deficiencies	Toxicity Risk	Complementary Element	Symptoms of Deficiency
	20 grams			Phosphorus Vitamin D	• Abnormal bone growth in young animals • Weakened bone structure in older animals • Muscular weakness and cramping • Cardiac irregularities • Thumps
	20-25 grams				
	25-30 grams				
	40 grams				
	40 grams				

Supplementation Recommendation

USE / DIET					
	0-10 grams	0-10 grams	0-15 grams	0-20 grams	0-20 grams
	0-12 grams	0-12 grams	0-17 grams	0-23 grams	0-23 grams
	0	0	0	0	0
	0	0	0	0	0
	0	0	0	0	0
	0	0	0	0	0

Sources

Calcium is available in all hays and grains, hays being higher than grains. Alfalfa hay is very high at 1.24%. Grass hays vary widely in their calcium content, from as low as around 0.2% to 0.4 to 0.45%.; grains being generally 0.1% or lower.

Functions

Calcium is required for the growth and maintenance of strong, normal bones and teeth. Calcium is also of great importance to normal functioning of the heart muscle and skeletal muscles and in sending nerve impulses. Calcium is also important to blood clotting and the release of some hormones.

Supplementation

Calcium supplementation will be needed when feeding some grass hays (hays with calcium content of less than 0.4%). These hays include Bermuda grass, late stages of Bahia grass, late stages of Kentucky bluegrass, orchard grass and brome grass. Pregnant and lactating mares, as well as growing horses, have very different calcium requirements from mature horses and you should seek expert advice regarding your diet and need for supplementation.

Deficiencies

Deficiencies are likely on all unsupplemented diets for late pregnancy and in young growing animals. Mature animals on some grass hays may also be deficient but do not necessarily show any symptoms. Symptoms of low calcium may also occur when calcium intakes are actually too high. What happens here is that the body recognizes the excessive calcium and takes measures to decrease absorption of the mineral. If a sudden drain on calcium occurs (e.g., a mare who begins making milk or a horse that has exercised heavily), the horse may be unable to produce enough calcium quickly enough. Trembling, excitability, muscular cramping and irregularities of the heart beat may result. Thumps is an interesting manifestation of low calcium that involves the diaphragm—the muscle that moves the chest up and down when the horse breathes. The diaphragm will contract forcefully with each heart beat, resulting in a quick, jerking movement that is visible just behind the horse's ribs. Horses on endurance rides often have increased needs for calcium.

Toxicity

Adult horses can tolerate calcium intakes of as high as 6 times the required amount if adequate phosphorus is being consumed. However, high intakes probably increase the risk of symptoms developing when a sudden calcium drain is applied (see Deficiencies). Excess calcium can also interfere with magnesium absorption. Excesses of calcium (or any mineral) and/or deviations from the ideal ratio of calcium to phosphorus should always be avoided in growing animals.

Interactions

Adequate vitamin D is required for absorption of calcium. Calcium absorption may also be decreased if phosphorus intake is not adequate. Inadequate protein in the diet depresses calcium absorption.

Indications

Calcium supplementation is indicated when feeding some grass hays in all pregnant, lactating and growing animals. In endurance horses, electrolyte supplements taken during the ride (as well as before and after) should contain generous amounts of calcium.

COBALT

TRACE MINERAL

Probability of Deficiency Per Use	RDA Per Use	Severity of Deficiencies	Toxicity Risk	Complementary Element	Symptoms of Deficiency
	·············· 0			Vitamin B12	• None known for horses (anemia in other species)
	·············· 0				
	·············· 0				
	·············· 0				
	·············· 0				

Supplementation Recommendation

USE / DIET					
	0	0	0	0	0
	0	0	0	0	0
	0	0	0	0	0
	0	0	0	0	0
	0	0	0	0	0
	0	0	0	0	0

Sources

The horse obtains cobalt from grasses, hays and grains.

Functions

Cobalt is used by the microorganisms of the intestinal tract to manufacture vitamin B12. This synthesized B12 is then absorbed by the horse's intestine.

Supplementation

No minimal requirement in the diet or recommended supplementation level has been set for horses. Horses do extremely well on pastures with very low cobalt levels without developing B12 deficiency.

Deficiencies

No deficiencies related to cobalt intake have been described in horses.

Toxicity

Unknown, unlikely to occur under natural circumstances.

Interactions

Cobalt is used by intestinal organisms to manufacture B12, which complements the level obtained from the diet.

Indications

None.

COPPER

TRACE MINERAL

Probability of Deficiency Per Use	RDA Per Use	Severity of Deficiencies	Toxicity Risk	Complementary Element	Symptoms of Deficiency
	50 mg			Vitamin C Protein Sulfur containing amino acids	• Bone and joint disease • Tendon and ligament problems • Poor hoof quality • Anemia • Discolored hair coat
	50-75 mg				
	75-100 mg				
	100-200 mg				
	100-200 mg				

Supplementation Recommendation

USE / DIET					
	50 mg	50-75 mg	75-100	100-200	100-200
	50 mg	50-75 mg	75-100	100-200	100-200
	50 mg	50-75 mg	75-100	100-200	100-200
	50 mg	50-75 mg	75-100	100-200	100-200
	50 mg	50-75 mg	75-100	100-200	100-200
	50 mg	50-75 mg	75-100	100-200	100-200

Sources

Hays and brans contain 10-20 ppm (mg/kg) of copper and grains less than 10 ppm. Molasses, the sweetener in sweet feeds, is an excellent source of copper, containing in excess of 80 ppm. Complete feeds are usually supplemented with copper.

Functions

Trace minerals are minerals that are essential to health but required in very small amounts. Copper is required for the production of normal connective tissues, including tendons and ligaments and the framework of the bones and cartilage lining joints. Inadequate copper intake in mares and foals has been implicated in the likelihood of developmental bone disease, including OCD—osteochondrosis dessicans—a debilitating joint disease that can ruin a horse's athletic career. Copper is also required for the mobilization of stored iron, needed to make red blood cells and prevent anemia. Copper is needed for the normal

Functions — continued

production of skin and coat pigments. Copper is also needed for normal development and function of the brain. Copper also has an important function in the activity of the antioxidant enzyme superoxide dismutase and other antioxidant functions. Exposure to drugs, chemicals, preservatives and inhaled impurities in the air may generate substances called free radicals. These electrically imbalanced molecules attack normal body tissues to steal an electrical charge that will restore them to a normal neutral state. The damaged cell in turn attacks its neighbors, setting up a chain reaction of cellular damage. A healthy horse constantly fights off invasion by a large variety of bacteria, viruses and other organisms. When the immune system carries out this function, a normal waste product is the generation of free radicals, potentially as damaging as any outside threat. Free radicals may be responsible for many of the familiar and uncomfortable symptoms of infection/inflammation (pain, swelling) and such things as sore throat, runny nose and other symptoms of viral infections. Exercise also results in free radical production that may cause muscle aching and fatigue. Antioxidants are substances present in the diet and manufactured by the body whose function is to neutralize free radicals before they can damage normal body tissues.

Supplementation

Increased intake of antioxidants may be called for in the case of infections, wounds, injuries such as sprains and strains, stressful situations such as shipping or change in environment, living in a polluted environment and, especially, exercise. With regard to copper's other important functions, the NRC has been reluctant to accept studies which suggest that copper levels required to prevent the development of bone and joint problems may actually be 4 to 5 times higher than the current NRC recommendations. However, arthritis, OCD and tendon/ligament problems are very widespread and significant problems in horses. The dosages for supplementation listed above strike a middle ground between current NRC estimates and possible greater needs. Since copper supplementation has a wide margin of safety, there is potentially much to gain by supplementing copper.

Deficiencies

General signs of less than optimal antioxidant intake include poor stress tolerance, exercise related problems including subpar performance, frequent infections, poor wound healing. Over the long term, suboptimal antioxidant intake is believed to contribute to premature aging, cancer and health problems such as heart disease and diseases of the blood vessels. Interestingly, inadequate copper levels have been implicated as a cause of rupture of the major artery of the uterus in older pregnant mares. Copper has also been implicated by research data in the development of bone and joint abnormalities such as OCD. Copper deficiency may also play a role in the development of arthritis, tendon and ligament problems, as well as poor hoof quality.

Toxicity

Studies in horses have shown copper to be very well tolerated even in high amounts. The estimated upper limit of safety is listed as 800 ppm, which would be 4,000 mg for a horse at maintenance. High copper intake may interfere with zinc absorption. Supplement to keep ratio of zinc:copper at about 3:1.

Interactions

Copper and zinc can probably be used interchangeably in the superoxide dismutase antioxidant system. Copper and zinc compete for absorption although this does not seem to be as signficant a problem in horses as it is in other species. A copper:zinc ratio of 3:1 is used in most equine supplements and commercial feeds to avoid any potential competition, but you should be more concerned with the total amounts of each in the diet than with their ratio to each other. A copper:zinc ratio of 3:1 is usually advised. Absorption problems can be avoided entirely by using a chelated (attached to a protein) form of copper instead of a copper in mineral form.

Indications

Because of copper's important role in the development, repair and maintenance of bones, joints, tendons, ligaments and hooves, supplementation is likely to be of benefit to any pregnant, growing or physically active horse. Its antioxidant functions and association with blood vessel disease make copper a wise mineral to supplement in any horse.

CYANOCOBALAMIN – VITAMIN B12

B VITAMIN

Probability of Deficiency Per Use	RDA Per Use	Severity of Deficiencies	Toxicity Risk	Complementary Element	Symptoms of Deficiency
	0			Folic acid Cobalt	• None known in the horse
	0				
	0				
	0				
	0				

Supplementation Recommendation

USE / DIET					
	0	0	0	0	0
	0	0	0	0	0
	0	0	0	0	0
	0	0	0	0	0
	0	0	0	0	0
	0	0	0	0	0

Sources

The horse receives B vitamins in his diet from such things as grains, brans and yeast. In addition, the abundant microorganism population in the horse's intestinal tract manufactures B vitamins and it is believed the horse can absorb a certain amount from this source. In the case of this special vitamin, which is not available from a vegetarian diet, the horse's source is solely the microorganisms in the intestine.

Functions

The B vitamins play essential roles in virtually every organ and every cell of the horse's body. B12 is critical to the function of every rapidly dividing cell type, to the formation of normal red blood cells (acting with folic acid) and in creation of the protective sheath or covering which surrounds all nerves.

Supplementation

B vitamins belong to the group called water soluble vitamins. This means they float freely through the fluids of the body and are not stored in any body tissues. Once these vitamins are absorbed into the blood, they circulate and are taken in by cells that need them or eliminated in the urine. Because of this rapid elimination, the horse must take in B vitamins on a daily basis. However, a small amount of B12 can be stored in the horse's tissues.

Deficiencies

Full blown, life-threatening deficiency symptoms for B vitamins are not likely to develop in horses on normal diets. Even if a horse is not eating, the organisms in the intestinal tract can manufacture these vitamins and some will be absorbed by the horse. However, horses can develop symptoms of inadequate amounts of some B vitamins. In the case of B12, there is no evidence to suggest B12 deficiencies ever occur, even on diets extremely low in the cobalt needed for the intestinal microorganisms to make this vitamin.

Toxicity

Toxicity with B vitamins administered in the feed is virtually nonexistent since the kidneys will eliminate any excess very efficiently. B12, even in large amounts, has not been reported to cause any toxicity in any species.

Interactions

Many of the B vitamins work in conjunction so it is usually advised that if you supplement with one B vitamin you should also supplement the others. However, B12 may be the one exception to this rule since the chance of deficiency appears to be extremely low. Adequate cobalt in the diet is necessary to maintain the horse's B12 status but the amount needed appears to be very low compared to other species (about 25 mg/day for horses at maintenance, 50 mg/day for heavy work).

Indications

In general, B vitamin supplementation is indicated in any horse that is not eating normally (or not eating at all), in horses that are heavily stressed for any reason (age, surgery, injury, infection, shipping, etc.) and in horses being heavily exercised. B12 supplementation should not need to be a major consideration, except in horses with prolonged periods of severe diarrhea or prolonged periods off feed with intravenous fluid nutrition (e.g., following a colic surgery).

FLUORINE

TRACE MINERAL

Probability of Deficiency Per Use	RDA Per Use	Severity of Deficiencies	Toxicity Risk	Complementary Element	Symptoms of Deficiency
	 0			None	• None for horses
	 0				
	 0				
	 0				
	 0				

Supplementation Recommendation

USE / DIET					
	0	0	0	0	0
	0	0	0	0	0
	0	0	0	0	0
	0	0	0	0	0
	0	0	0	0	0
	0	0	0	0	0

Sources

Uncontaminated forages contain 2 to 16 ppm of fluorine; grains 1 to 3 ppm.

Functions

Small amounts of fluorine are needed for normal hardness of bones and teeth.

Supplementation

None advised.

Deficiencies

Deficiency is not known to exist in the horse. The major manifestation of inadequate fluorine in other species is tooth decay. Complete deficiency could be expected to cause abnormalities of bone.

Toxicity

Situations leading to ingestions of large amounts of fluorine are unlikely to occur. Overdose in people can lead to bone damage, kidney damage, nerve and muscle disease and discoloration of the teeth.

Interactions

None.

Indications

None.

FOLIC ACID

B VITAMIN

Probability of Deficiency Per Use	RDA Per Use	Severity of Deficiencies	Toxicity Risk	Complementary Element	Symptoms of Deficiency
	0			B12	• Mild anemia • Poor general condition • Poor performance
	0				
	0				
	0				
	0				

Supplementation Recommendation

USE / DIET					
	0	0	0	20mg	20mg
	0	0	0	20mg	20mg
	0	0	0	20mg	20mg
	0	0	0	20mg	20mg
	0	0	0	20mg	20mg
	0	0	0	20mg	20mg

Sources

Grains, brans and yeast provide the B vitamins in a horse's diet. In addition, the abundant microorganism population in the horse's intestinal tract can manufacture B vitamins and horses may absorb a certain amount from this source. Studies of folic acid confirm this vitamin is synthesized in the intestines. The horse consumes folic acid in the form of natural folacin compounds. The concentration of these substances is much higher in fresh grasses (1,630 micrograms/kg) than hays (670) or oats (210). Wheat bran is approximately the same as hays (650 mg).

Functions

The B vitamins play essential roles in virtually every organ and every cell of the horse's body. Folic acid, working in conjunction with vitamin B12, is especially important to the normal production of red blood cells/ prevention of anemia. Folic acid is also essential for normal protein metabolism and in the duplication of cells.

Supplementation

B vitamins are water soluble. They float freely through the fluids of the body and are not stored in any body tissues. Once these vitamins are absorbed into the blood, they circulate and are taken in by cells that need them or eliminated in the urine. Because of this rapid elimination, the horse must take B vitamins on a daily basis. Stabled horses take in far less folic acid than horses on pasture, as much as 9 to 10 times less.

There is no NRC recommended daily intake for folic acid in horses although they clearly have times when supplementation is advisable (see below). A dosage of about 1.5 mg/day would be the equivalent of a human dose. However, horses are thought not to efficiently absorb folic acid in the form of supplements. One study reported benefit from 20 mg per day of folic acid.

Deficiencies

Horses on normal diets are not likely to develop full blown, life-threatening deficiency symptoms for B vitamins. Even if a horse is not eating, the organisms in the intestinal tract can manufacture these vitamins and some will be absorbed by the horse. However, the horse can develop symptoms of inadequate amounts of B vitamins. Specifically, studies have clearly demonstrated that training/heavy exercise result in decreased serum folic acid levels. A confirmed case of mild anemia and poor body condition, poor performance has been reported in a horse that showed improvement in all three problems when he was supplemented with 20 mg of folate per day. In other species, folic acid deficiency in pregnant women has been linked to the development of spinal cord abnormalities in the fetus. This has not been seen in the horse. The classic picture of large pale red blood cells and anemia with folic acid deficiency has not been seen in horses, probably because their chances of B12 deficiency are extremely low. (See vitamin B12)

Toxicity

Toxicity with B vitamins administered in the feed is virtually nonexistent since the kidneys will eliminate any excess very efficiently. Extremely large amounts given intravenously have produced seizures in other species and intakes of 20 times the recommended daily allowance in people has been suspected to cause intestinal upset.

Interactions

Many of the B vitamins work together, so it is usually advised that if you supplement with one B vitamin you should also supplement the others. As mentioned, folic acid works hand in hand with B12 in maintaining normal red blood cell counts.

Indications

B vitamin supplementation may be indicated in any horse not eating normally (or not eating at all), in horses that are heavily stressed for any reason (age, surgery, injury, infection, shipping, etc.) and in horses being heavily exercised. Folic acid is one vitamin where we actually have confirmation that heavy exercise causes levels to drop. Horses denied access to fresh grass could also benefit from folic acid supplementation.

IODINE

TRACE MATERIAL

Probability of Deficiency Per Use	RDA Per Use	Severity of Deficiencies	Toxicity Risk	Complementary Element	Symptoms of Deficiency
	0.5-3.0 mg			Protein Zinc Selenium	Symptoms of hypothyroidism including: • abnormal growth • weakness • poor appetite • overweight or muscle loss • poor performance and easy fatigue • muscle cramping and pain, tying-up • decreased fertility • enlargement of thyroid gland • weak foals • leg deformities
	0.5-3.0 mg				
	0.75-4.5 mg				
	1.0-6.0 mg				
	1.0-6.0 mg				

Supplementation Recommendation

USE / DIET	(pasture)	(riding)	(jumping)	(racing)	(harness)
(grass)	0 mg	0 mg	0 mg	1 mg	1 mg
(grass + oats)	0 mg	0 mg	0 mg	1 mg	1 mg
(clover)	0 mg	0 mg	0-3 mg	0-4.5 mg	0-4.5 mg
(clover + oats)	0 mg	0 mg	0-3.5 mg	0-5 mg	0-5 mg
(clover + grass)	0 mg	0 mg	0 mg	1 mg	1 mg
(grass + clover + oats)	0 mg	0 mg	0-2.5 mg	0-4 mg	0-4 mg

Sources

Iodine is present in all hays and grains at levels which will reflect the local soil levels. Using NRC average figures, grass hay is a richer source than alfalfa and both considerably higher than grains.

Functions

Iodine is required for the normal production of thyroid hormone. Thyroid hormone plays central roles in the regulation of growth, regulation of energy metabolism and determination of the sensitivity of muscles/speed at which they contract.

Supplementation

The usual method of supplementing iodine is to provide the horse with a trace mineral salt block (the brown ones). These provide 70 mg of iodine/kg. A 3.5 pound block contains about 111 mg of iodine. If the horse eats 1 oz of the block per day, he will be taking in 2 mg of iodine. This would be adequate for all but the most stressful situations.

Deficiencies

Deficiency of iodine is most likely to result in impaired fertility, weak newborns and exercise related problems. Poor appetite may also develop in stressed horses and prevent them from showing the weight gain associated with hypothyroidism under other circumstances. Hypothyroidism has been shown to cause muscular weakness, pain, cramping and tying-up in heavily exercised horses. It is not known if this is related directly to iodine deficiency or some other mechanism, but it makes sense to guarantee that heavily worked horses receive adequate iodine. Hypothyroidism has been suspected in overweight, "lazy" warmblooded breeds and in fat ponies with a tendency to founder. However, abnormal thyroid hormone levels have not been confirmed. If T4 levels are normal, a possible explanation would be decreased conversion of T4 to the active T3 hormone, which could be related to selenium or zinc deficiency.

Toxicity

Iodine toxicity has been seen in horses fed excessive high iodine supplements. These include products based on kelp or other seaweeds. Any supplements of marine product bases could do the same thing. Symptoms include complete hair loss and the development of goiter (a swelling in the neck from an enlarged thyroid gland).

Interactions

Normal thyroid hormone production also requires a high quality protein diet and adequate levels of selenium and zinc, the latter for converting thyroid hormone to its active form.

Indications

Usual diets with free access to trace mineral salt blocks will provide adequate iodine for most circumstances. Heavily stressed/exercised horses and pregnant mares may need additional supplementation of 1 to 2 mg of iodine per day for maximal intake.

IRON

TRACE MINERAL

Probability of Deficiency Per Use	RDA Per Use	Severity of Deficiencies	Toxicity Risk	Complementary Element	Symptoms of Deficiency
	250mg			Riboflavin Pyridoxine Vitamin C	• Anemia • Susceptibility to infections
	250-375mg				
	375mg				
	500mg				
	500mg				

Supplementation Recommendation

USE / DIET					
(hay)	0	0	0	0	0
(hay + oats)	0	0	0	0	0
(clover)	0	0	0	0	0
(clover + oats)	0	0	0	0	0
(hay + clover)	0	0	0	0	0
(hay + clover + oats)	0	0	0	0	0

Sources

Virtually every commonly fed hay and grain contains iron in excess of the level required by horses. Corn is the exception (at just over 30 ppm compared to the minimal required 50 ppm). However, grasses are so generously supplied this is easily balanced by the hay portion of the ration.

Functions

Iron is needed for the production of the red blood cells that carry oxygen and the white blood cells that fight infection.

Supplementation

Normal diets supply more than enough iron for horses of all ages and uses.

Deficiencies

Iron deficiency is virtually non-existent except in horses that have sustained a very large loss of blood. The horse has abundant iron stored in his bone marrow and also carries a very large reserve of red blood cells in his spleen.

Toxicity

Iron is definitely toxic in large amounts. Foals given a paste supplement to promote growth of intestinal organisms have died from the iron in the product. Long term feeding of iron-containing supplements can cause chronic toxicity with toxic iron levels in the kidneys and liver. The mane, tail and coat may take on a rusty color. High iron levels interfere with the absorption of important trace minerals and can produce a wide variety of problems, depending on which are involved (see copper, manganese and zinc). Excessive iron also predisposes horses to infections by interfering with immune function. Chronic fatigue, arthritis, low level abdominal pain, impaired fertility, damage to the heart muscle and liver are all common symptoms of iron overload in other species and can just as easily occur in horses. Injectable iron is the most dangerous but many "blood tonics" also contain dangerous levels of iron. Worst of all, there is no treatment for iron overload except to remove blood at regular intervals. Certainly this is not what you are after when you feed those high iron supplements!

Interactions

High iron interferes with absorption of copper and zinc. Excessive calcium supplementation may compete with iron for absorption. Note that alfalfa hay has very high calcium levels, but also contains an extremely high level of iron. The presence of high iron levels in some supplements can inactivate vitamin C and vitamin E.

Indications

None, except anemia following a massive blood loss. Even allowing for increased iron needs in high performance horses, the diet will easily meet those needs.

LYSINE

AMINO ACID

Probability of Deficiency Per Use	RDA Per Use	Severity of Deficiencies	Toxicity Risk	Complementary Element	Symptoms of Deficiency
	23 grams			Balanced protein B6	• Poor quality coat and hooves
	29 grams				• Abnormal growth
	34 grams				• Abnormal bone formation
	46 grams				• Poor tolerance to exercise, inability to build muscle
	5 grams				

Supplementation Recommendation

USE / DIET					
grass	5.5 grams	5.5 grams	7.75 grams	11 grams	11 grams
grass + grain	4.6 grams	1.4 grams	2.5 grams	9.8 grams	9.8 grams
alfalfa	0	0	0	0	0
alfalfa + grain	0	0	0	0	0
grass + alfalfa	0	0	0	0	0
grass + alfalfa + grain	0	0	0	0	0

Sources

Protein is available from all common feeds. Fresh grasses and alfalfa hay contain the most, followed by grains and grass hays. The horses body can manufacture some amino acids.

Functions

Amino acids are known as the building blocks of protein. Essential amino acids must be present in the diet because the body cannot make them, while nonessential amino acids can be manufactured by the horse. The essential amino acids for people are: threonine, lysine, valine, leucine, methionine, isoleucine, tryptophan, phenylalanine and histidine. In the horse, only lysine has been demonstrated to be essential. There is considerable interest currently in methionine and threonine but not enough information has been compiled to advise on required dietary levels. The amino acids leucine, isoleucine and valine are of interest to those who work with high performance horses. Lysine is important to the proper use of all amino acids.

Supplementation

Diets based on grass hays or grass hay and grains may not meet even NRC recommendations for total protein and lysine content unless hays are very high quality and cut early in their growth. Grains may also fall below needed levels depending upon where they were grown and the ratio of hull to kernel. All corn is low in lysine. You can use a supplement containing additional lysine, lysine alone or a high quality protein supplement such as milk protein (e.g., dried skim milk), whey or combined soybean/milk protein products.

Deficiencies

Deficiency of any of the essential amino acids will lead to poor growth and poor tolerance to stress. Specific symptoms for an isolated amino acid will depend upon those organ systems where it is particularly important. Lysine deficiencies on grass hay are not large enough to cause any major health problems but you may have lower quality coat and hooves, delayed wound healing and poor tolerance to the stress of exercise with little ability to build muscle, poor red blood counts and decreased fertility.

Toxicity

Large doses may decrease absorption of the amino acid arginine and can make the kidneys more sensitive to damage from the aminoglycocide antibiotics (i.e., the mycins).

Interactions

There are many complex interactions between amino acids and other amino acids. The proper functioning of amino acids is also closely tied to adequate energy content in the diet and normal vitamin and mineral levels. Excessive lysine may create an arginine deficiency.

Indications

To bring the lysine level of any inadequate diet up to NRC recommendations.

MAGNESIUM

MAJOR MINERAL

Probability of Deficiency Per Use	RDA Per Use	Severity of Deficiencies	Toxicity Risk	Complementary Element	Symptoms of Deficiency
	6.75 grams			Potassium Calcium	• Muscular weakness
	8.56 grams				• Tying-up
	10.28 grams				• Skin hypersen–sitivity
	13.70 grams				• Irritability
	13.70 grams				• Thumps

Supplementation Recommendation

USE / DIET					
(grass)	0-4.37	0-5.47	0-6.55	0-8.74	0-8.74
(grass + oats)	0	0	0	0	0
(clover)	0-4	0-15	0-19	0-25.2	0-25.2
(clover + oats)	0-2	0-2.5	0-3.6	0-4.1	0-4.1
(grass + clover)	0-9	0-11.3	0-13.5	0-18	0-18
(grass + clover + oats)	0-1.7	0-2.2	0-2.6	0-3.4	0-3.4

NOTE: These numbers refer only to the approximate amounts of additional magnesium needed to bring the ratio of calcium:magnesium into the range of 2.5:1 (see discussions below). *All diets are adequate in terms of total magnesium ingestion. Routine supplementation is not required unless symptoms suggestive of low magnesium are noted.*

Sources

Magnesium is found in all common hays and grains, being highest in alfalfa, seed meals and brans.

Functions

Magnesium is critical to the normal functioning of heart muscle, skeletal muscle and the nervous system. It is also a component of over 300 enzyme systems throughout the body, where it is important in all energy generating systems.

Supplementation

Although the total magnesium intake is adequate for all diets and uses, there is a potential problem since the amount of calcium being consumed is often in excess of the ideal 2:1 ratio reported for other species and the 2.5:1 average ratio found in natural grasses. At higher ratios, calcium can compete with magnesium for absorption to the point that inadequate magnesium is actually absorbed. However, the picture is further complicated by the observation in other animals and people that low magnesium levels in the feed may be associated with better than normal absorption. This could explain why many animals do not develop problems related to calcium excess.

Blood tests for magnesium are normal to low in deficiency states. However, blood magnesium is a poor indicator of magnesium status inside the muscle cells and a syndrome of elevated muscle enzymes with muscle pain, cramping and even tying-up has been seen. The only truly reliable way to determine magnesium status inside the cells is by sophisticated testing of biopsy samples or hair mineral analysis.

If your horse is on alfalfa hay or mixed alfalfa/grass hay and experiencing muscle problems, it may help to supplement with magnesium. Supplementation with 2 to 3 grams/day is usually adequate to control the symptoms but much higher amounts are needed to actually result in a rise in blood magnesium levels.

Deficiencies

As noted above, magnesium deficiency is unlikely if all you look at is the total amount of magnesium being consumed. However, horses receiving excessive calcium in their diets can develop magnesium deficiency symptoms, probably caused by competition for absorption from the excessive calcium. The more hay the horse receives in comparison to grain, the greater the risk. In these cases, 2 to 3 grams of magnesium per day may be sufficient to correct the outward symptoms but much higher amounts would be needed to bring calcium and phosphorus back into a correct ratio.

Toxicity

No known toxicity for oral magnesium. Can cause diarrhea in other species. Potential for some sedative/depressive effect at high doses.

Interactions

High calcium in diet may block absorption of magnesium.

Indications

Horses showing symptoms of muscular problems, hypersensitivity/irritability on diets that may have adequate total magnesium but excessive amounts of calcium. Endurance exercise may also lead to higher magnesium losses than normal, aggravating the problem and leading to problems during competition such as those above and/or thumps.

MANGANESE

TRACE MINERAL

Probability of Deficiency Per Use	RDA Per Use	Severity of Deficiencies	Toxicity Risk	Complementary Element	Symptoms of Deficiency
	..........200mg			Vitamin C Copper	• Abnormal development of bones and joints
	..........200-300mg				• Impaired ability to make/repair joint cartilage
	..........300mg				• Abnormalities of hair, skin and hooves
	..........400mg				• Impaired fat and carbohydrate metabolism
	....400mg				

Supplementation Recommendation

USE DIET					
	0	0	0	0	0
	0	0	0-100 mg	100 mg	100 mg
	100 mg	100-150mg	150 mg	200 mg	200 mg
	100 mg	100 mg	150 mg	200 mg	200 mg
	0	0	0	0	0
	0	0	0-100 mg	100 mg	100 mg

Sources

Of the commonly fed concentrates, oats and soybean fall just below the NRC recommendation of 40 ppm (mg/kg) while barley and corn are extremely low in manganese. Alfalfa hay is also lower than the NRC recommended level and grass hays vary widely but are generally adequate to good sources.

Functions

Manganese functions in the normal metabolism of sugar and fat; to make the cholesterol backbone of important hormones and for cells to be able to duplicate their genetic material (DNA) and divide. It is essential for the production of chondroitin sulfates—important in maintaining normal joint cartilage and repairing cartilage.

Supplementation

Hay-only diets of grass hay or mixed grass and alfalfa hay will provide adequate manganese. Use of either of these hays with a grain ration that is primarily or entirely oats will also probably be adequate. However, alfalfa-based, hay only or hay and grain diets will be too low in manganese to meet even NRC minimum recommendations. Suggested dosages above also take into consideration what are probably increased needs of high performance horses with more regular wear and tear on their joints.

Deficiencies

As described above, many common diets are likely to be deficient in manganese. The manganese recommendations of the NRC are not based on studies in horses but taken from data for other animals. Need for a nutrient rises when a condition exists that calls for more of it. Certainly the heavy use of joints in performance horses and great number of horses that have problems with arthritis, not to mention the problem of developmental bone and joint disease in young horses, make a strong case for efforts to guarantee adequate intake of this and other joint related nutrients. Dosages above take into consideration both likely levels in the diet and likely increased needs with exercise.

Toxicity

Toxicity has not been reported for horses or other farm animals and is not likely to occur under natural conditions.

Interactions

A high intake of iron depresses absorption of manganese.

Indications

As above, pregnant, growing and exercising horses, as well as horses with arthritis need adequate manganese. Manganese levels in many common diets are likely to be deficient.

NIACIN

B VITAMIN

Probability of Deficiency Per Use	RDA Per Use	Severity of Deficiencies	Toxicity Risk	Complementary Element	Symptoms of Deficiency
	0			Tryptophan Other B Vitamins	• Nervousness
	0				• Excitability
	0				• Jumpiness
	0				• Inflammation and ulcers on tongue and in mouth
	0				• Fatigue
					• Poor appetite

Supplementation Recommendation

USE / DIET					
(grass)	0	0	50 mg	100 mg	100 mg
(grass + bucket)	0	0	0	50 mg	50 mg
(clover)	0	0	50 mg	100 mg	100 mg
(clover + bucket)	0	0	0	50 mg	50 mg
(grass + clover)	0	0	50 mg	100 mg	100 mg
(grass + clover + bucket)	0	0	0	50 mg	50 mg

Sources

The B vitamins in a horse's diet largely come from grains, brans and yeast. In addition, the abundant microorganism population in the horse's intestinal tract can make B vitamins and it is believed the horse can absorb a certain amount from this source. Studies of thiamine levels in intestinal contents throughout the horse's digestive tract show concentrations much higher than those present in the horse's diet. It is also believed that the horse can synthesize niacin from tryptophan (an amino acid) in his own body.

Functions

The B vitamins play essential roles in virtually every cell of the horse's body. Specifically, niacin is needed for proper energy production, for protein metabolism, metabolism of fatty acids and cholesterol, copying of DNA, reproduction of cells and in normal functioning of the nervous system.

Supplementation

The B vitamins are water soluble vitamins which float freely through the fluids of the body and are not stored in any body tissues. Once these vitamins are absorbed into the blood, they circulate and are taken in by cells that need them or are eliminated in the urine. Because of this rapid elimination, the horse must take in needed B vitamins on a daily basis. There is no official NRC recommendation for intake but it has been noted that a horse taking in the 100 mg/1,000 lb will be excreting niacin in his urine, indicating there is excess niacin in the body. However, heavily worked horses, on high grain intakes, may require more than the available supply which is why supplementation is recommended for some activity levels.

Deficiencies

In a horse's normal diet, full blown, life-threatening deficiency symptoms of B vitamins are not likely to develop. Even if a horse is not eating, the organisms in the intestinal tract can manufacture these vitamins and some will be absorbed by the horse. However, a horse can develop symptoms of inadequate amounts of B vitamins. Specifically, the nervous, irritable and difficult-to-handle horse may exhibit symptoms of B vitamin inadequacy, including niacin. Classic deficiency states with ulcerations in the tongue and mouth have not been seen in the horse.

Toxicity

Toxicity with B vitamins administered in the feed is virtually nonexistent since the kidneys will eliminate any excess. Administration of niacin in high doses can cause side effects of a burning and itching sensation on the skin, with flushing of the skin. In the horse, this would most likely be manifested as rubbing against objects and severe agitation.

Interactions

Many of the B vitamins work in conjunction so it is usually advised that if you are supplementing with one B vitamin you should also supplement the others.

Indications

In general, B vitamin supplementation is indicated in any horse that is not eating normally (or not eating at all), horses with digestive tract problems, in horses that are heavily stressed for any reason (age, surgery, injury, infection, shipping, etc.) and in horses that are being heavily exercised. In addition, horses that are jumpy, nervous and difficult to work around may manifest symptoms of B vitamin inadequacy, including niacin. Other possible symptoms include diarrhea and decreased exercise tolerance/fatigue.

PANTOTHENIC ACID

B VITAMIN

Probability of Deficiency Per Use	RDA Per Use	Severity of Deficiencies	Toxicity Risk	Complementary Element	Symptoms of Deficiency
	0			All B vitamins	• Poor energy level
	0				• Possible muscular symptoms
	0				• Possible burning feet
	0				• Nervousness, irritability
	0				

Supplementation Recommendation

USE / DIET					
	0	0	20 - 40 mg	100-200 mg	100-200 mg
	0	0	20- 40 mg	100-200 mg	100-200 mg
	0	0	20-40 mg	100-200 mg	100-200 mg
	0	0	20-40 mg	100-200 mg	100-200 mg
	0	0	20-40 mg	100-200 mg	100-200 mg
	0	0	20-40 mg	100-200 mg	100-200 mg

Sources

The horse gets plenty of B vitamins in his diet from such things as grains, brans and yeast. In addition, the abundant microorganism population in the horse's intestinal tract can make B vitamins, and it is believed the horse can absorb a certain amount from this source.

Studies that look at pantothenic acid in the intestinal fluid and compare it to the level found in the horse's diet appear to confirm significant synthesis in the intestinal tract.

Functions

The B vitamins play essential roles throughout the horse's body. This vitamin is extremely important in energy metabolism—fats and glucose, as well as in the manufacture of steroid hormones and chemicals involved in the transmission of nerve impulses. It is also required for adequate absorption and use of folic acid.

Supplementation

The B vitamins are water soluble. This means they float freely through the fluids of the body and are not stored in any tissues. Once absorbed into the blood, these vitamins circulate and are taken in by cells that need them or eliminated in the urine. Because of this rapid elimination, a horse should receive B vitamins on a daily basis. There is currently no NRC recommended minimal daily intake for pantothenic acid. Supplementation is probably not beneficial with the exception of horses with intestinal tract problems, under stress and/or in heavy work.

Deficiencies

It is not likely that horses on a normal diet will develop full blown, life-threatening deficiency symptoms for B vitamins. Even if a horse is not eating, organisms in the intestinal tract can make these vitamins and some will be absorbed by the horse. However, the horse can develop symptoms of inadequate amounts of B vitamins. Specifically, pantothenic acid deficiency would be expected to result in defective energy production, poor stress tolerance and nervousness. Interestingly, a consistently reported symptom observed in people with inadequate pantothenic acid intake is feet that feel as if they are burning.

Toxicity

Toxicity with B vitamins administered in the feed is virtually nonexistent since the kidneys will eliminate any excess very efficiently.

Interactions

Many B vitamins work together so it is usually advised that if you supplement one B vitamin you should also supplement the others, especially if efficient energy production for performance is your goal. Pregnant mares should receive supplemental pantothenic acid and folic acid to prevent birth defects (50 mg/day is recommended).

Indications

In general, B vitamin supplementation is indicated in any horse that is not eating normally (or not eating at all), in horses with intestinal tract problems, in horses that are heavily stressed for any reason (age, surgery, injury, infection, shipping, etc.) and in horses that are being heavily exercised.

PHOSPHORUS

MAJOR MINERAL

Probability of Deficiency Per Use	RDA Per Use	Severity of Deficiencies	Toxicity Risk	Complementary Element	Symptoms of Deficiency
	14 grams			Calcium	• Abnormal bone growth in young animals
	18 grams				• Weakened bone in older animals
	21 grams				
	29 grams				
	29 grams				

Supplementation Recommendation

USE / DIET					
	2 grams	0	0	5 grams	5 grams
	0	0	0	0	0
	3.0 grams	1.5 grams	0	7 grams	7 grams
	2.5 grams	0	0	4 grams	4 grams
	2.5 grams	.75 grams	0	6 grams	6 grams
	1 grams	0	0	3.25 grams	3.25 grams

Sources

Phosphorus is present in all hays and grains. A value of 0.24% was used for grass hays as an average although costal hays have less (under 0.2%). Alfalfa is 0.22% phosphorus and oats 0.34%. Bran is an excellent source of phosphorus, containing 1.13%. One pound of bran supplies 5 grams of phosphorus and very little calcium.

Functions

Phosphorus' major role in the body, at least in terms of amount used, is in the normal formation of bones. Adequate phosphorus is needed for calcium to be used properly. An ideal ratio of calcium:phosphorus is 1.2 to 1.5:1. However, the total amount of phosphorus is most important since excessive calcium can be handled if adequate phosphorus is being taken in. Phosphorus also is part of many high energy compounds in the body, required for all bodily functions (e.g., ATP and phosphocreatine). Phosphorus compounds are also important buffers or neutralizers of acid in the body, such as the lactic acid produced during burning of carbohydrates when animals are working hard.

Supplementation

Phosphorus intake is borderline to low for most uses of horses on hay only diets. The addition of grain with its higher phosphorus content helps to fill the gap but many horses, especially horses taking in limited amounts of feed and horses being heavily exercised, are still somewhat deficient. Pregnant, lactating and growing animals have special phosphorus needs and you should get expert advise for horses in these categories.

Deficiencies

The most obvious deficiencies occur in developing and growing horses deprived of adequate phosphorus (bone deformities). Inadequate phosphorus may also have a negative impact on performance horses. Short term loading with phosphorus has been proven to improve both endurance and speed in human athletes.

Toxicity

Slight overfeeding of phosphorus is not harmful but long term intake of high phosphorus causes a disease called nutritional secondary hyperparathyroidism. This disturbance in calcium and phosphorus metabolism results in a shifting lameness, then enlargement of the bones of the jaw and crest of the face.

Interactions

Calcium and phosphorus function together in the formation and maintenance of strong, healthy bones.

Indications

Phosphorus should be supplemented in any horse whose diet does not meet the minimal NRC requirements.

POTASSIUM

MAJOR MINERAL

Probability of Deficiency Per Use	RDA Per Use	Severity of Deficiencies	Toxicity Risk	Complementary Element	Symptoms of Deficiency
	25 grams			Chloride Sodium Calcium	• Muscular weakness and cramping • Cardiac irregularities • Thumps
	31.3 grams				
	37.4 grams				
	49.9 grams				
	49.9 grams				

Supplementation Recommendation

USE / DIET					
	0	0	0	0	0
	0	0	0	0	0
	0	0	0	0	0
	0	0	0	0	0
	0	0	0	0	0
	0	0	0	0	0

*Special needs call for this electrolyte. See Chapter 5, Nutrition and Performance.

Sources

Potassium is present in abundant supply in all feeds. Hays have the most, containing about 1.4% potassium, grains less and averaging about 0.4% potassium.

Functions

Potassium is the main mineral (electrolyte) found inside the cells. Like magnesium and calcium, it is important to muscle contraction in the heart and skeletal muscles. Potassium is also important in the transmission of nerve impulses.

Supplementation

Supplementation is not necessary under normal conditions. However, when the temperature is very high, losses in sweat will be great and supplementation may be necessary (usually about 4 grams per day of potassium chloride). Horses racing on Lasix also need extra potassium since this drug causes sudden and massive potassium losses in the urine that can result in muscular cramping, trembling and weakness. Horses with chronic muscle problems or those recovering from tying-up also may have greater potassium needs since the muscle damage causes potassium to leak out into the blood where the kidneys will get rid of it.

Deficiencies

As mentioned, potassium deficiency causes muscular weakness and cramping. It can also cause irregularities of the heart beat and thumps, a condition where the diaphragm (the muscle which moves the chest up and down when the horse breathes) contracts suddenly and violently with each heart beat, causing a visible jerking/thumping behind the horse's ribs.

Toxicity

Orally administered potassium has little risk of toxicity. This is because it is absorbed gradually, allowing the kidneys time to clear any excess before it can build up to dangerous levels in the blood.

Interactions

Potassium works with calcium and magnesium in the normal contraction of muscle cells.

Indications

Horses with muscle problems and horses exercising in the heat may require more potassium than the diet provides. Horses receiving the drug Lasix also have higher potassium requirements to prevent side effects.

PYRIDOXINE — VITAMIN B6

B VITAMIN

Probability of Deficiency Per Use	RDA Per Use	Severity of Deficiencies	Toxicity Risk	Complementary Element	Symptoms of Deficiency
	 0			Niacin Tryptophan	• Inability to properly use protein
	 0				• Behavior changes in mares with estrus (heat)
	 0				• Impaired ability to use glycogen, the major fuel for working muscles, causing decreased endurance, speed, weakness, cramping and possibly tying-up
	 0				• Nervousness, irritability
	 0				• Mouth and tongue ulcers

Supplementation Recommendation

USE / DIET					
(hay)	20-60 mg	40-120 mg	120-250 mg	250-300 mg	250-300 mg
(hay + oats)	20-60 mg	40-120 mg	120-250 mg	250-300 mg	250-300 mg
(pasture)	0	24 mg	60 mg	125-150 mg	125-150 mg
(pasture + oats)	0	24 mg	60 mg	125-150 mg	125-150 mg
(hay + pasture)	0-12 mg	24-60 mg	60-120 mg	60-120 mg	60-120 mg
(hay + pasture + oats)	0-12 mg	24-60 mg	60-120 mg	60-120 mg	60-120 mg

Sources

B vitamins in a horse's diet may be found in grains, brans and yeast. Also, the horse's intestinal tract has an abundant microorganism population that can manufacture B vitamins, which the horse can absorb. Studies have shown some pyridoxine is manufactured in the intestine, including the small intestine, in excess of dietary levels but the amount is relatively small. Information is sketchy regarding dietary levels of this vitamin but it is high in bran, moderately high in most unprocessed grains (not oats) and likely present in good amounts in fresh grasses but not grass hays.

Functions

B vitamins play essential roles almost everywhere in the horse's body. Pyridoxine is especially important to the use of dietary protein. It plays a role in the conversion of tryptophan into niacin. Of special importance to exercising horses is pyridoxine's critical function in the ability of the muscle cells to use their main source of energy—stored carbohydrate in the form of glycogen. Pyridoxine also assists in the production of neutrotransmitters, chemicals within the brain.

Supplementation

B vitamins are water soluble. Floating freely through the fluids of the body, they are not stored in any body tissues. Once these vitamins are absorbed into the blood, they circulate and are taken in by cells as needed them or eliminated in the urine. Because of this rapid elimination, the horse must take needed B vitamins daily. Requirements for pyridoxine will be higher in horses with a poor dietary intake, exercising horses and horses on a high protein diet.

Deficiencies

Full blown, life-threatening B vitamin deficiency symptoms are not likely to develop in horses on normal diets. Even if a horse is not eating, organisms in the intestinal tract make these vitamins and some will be absorbed by the horse. Horses can develop symptoms of inadequate amounts of B vitamins. Inadequate pyridoxine intake should be suspected in horses that have difficulty maintaining muscle mass, horses with poor exercise tolerance/performance and horses prone to muscle pain, cramping and tying-up. Pyridoxine deficiency would have the most severe consequences in horses that rely on anaerobic pathways of energy generation—i.e., horses required to work at maximal speed. Pyridoxine has also been associated with premenstrual syndrome in humans and could play a role in mares who exhibit behavior changes in their estrus (heat) cycles. Nervousness/irritability may be manifestations of inadequate pyridoxine in any horse because of its function in the production of brain chemicals.

Toxicity

Toxicity with B vitamins administered in the feed is usually virtually nonexistent since the kidneys efficiently eliminate any excess. However, pyridoxine is an exception. Excessive pyridoxine can cause severe nervous system problems including altered sensation, incoordination and even seizures. The precise toxic dose for horses is unknown. Extrapolating from information we have regarding toxicity in humans, use of pyridoxine in doses of over 600 mg per day for long periods could cause toxicity, at least in terms of abnormal sensation. Incoordination and convulsions would probably require doses of 2,000 to 6,000 mg/day.

Interactions

Many B vitamins work together, so it is generally recommended that if you supplement with one B vitamin you should also give the others. Since pyridoxine is required for conversion of tryptophan to niacin, it should be supplemented if tryptophan is being given. Pyridoxine needs are also closely tied to protein intake—increased protein increases the need for pyridoxine.

Indications

In general, B vitamin supplementation is indicated in any horse that is not eating normally (or not eating at all), in horses that are heavily stressed for any reason (age, surgery, injury, infection, shipping, etc.), in horses with chronic digestive problems and in horses being heavily exercised. Pyridoxine supplementation is strongly advised for exercising horses and horses with a history of muscle problems.

RIBOFLAVIN

B VITAMIN

Probability of Deficiency Per Use	RDA Per Use	Severity of Deficiencies	Toxicity Risk	Complementary Element	Symptoms of Deficiency
	20 mg			Niacin Tryptophan Pyridoxine Folic acid Vitamin K Thyroid hormone	• Most likely to occur in horses with high aerobic energy outputs, such as endurance horses and Thoroughbred race horses.
	20 mg				
	20 mg				
	20 mg				
	20 mg				

Supplementation Recommendation

USE DIET					
	0-20 mg	0-20 mg	0-20 mg	40-60 mg	40-60 mg
	0-20 mg	0-20 mg	0-20 mg	40-60 mg	40-60 mg
	0	0	0-20 mg	20-40 mg	20-40 mg
	0	0	0-20 mg	20-40 mg	20-40 mg
	0	0	0-20 mg	20-40 mg	20-40 mg
	0	0	0-20 mg	20-40 mg	20-40 mg

Sources

Grains, brans and yeast are a good source of B vitamins in a horse's diet. Fresh grasses, alfalfa hay (and meal) also supply riboflavin in amounts that exceed the recommended dietary levels, as does corn. Oats and grass hays are poor sources of riboflavin. The microorganism population found in the horse's intestinal tract can also make B vitamins, which horses can absorb. Studies have demonstrated riboflavin levels in the small intestine (where absorption is most likely) and large intestine that greatly exceed those found in the horse's diet.

Functions

B vitamins are essential to a horse's good health. Riboflavin is essential to the conversion of tryptophan into niacin and in the normal functioning of other B vitamins (see above) and vitamin K. Adequate thyroid hormone is needed for the conversion of riboflavin into its active form. Riboflavin plays critical roles in the generation of energy aerobically (in the presence of oxygen). Horses with high aerobic energy outputs, horses on high-fat diets or any concentrated energy source (grains) may require more riboflavin.

Supplementation

B vitamins, like other water soluble vitamins, float freely through the fluids of the body and are not stored in body tissues. Once these vitamins are absorbed into the blood, they circulate and are taken in by cells that need them or eliminated in the urine. Because of this rapid elimination, the horse must take needed B vitamins on a regular basis. Because of the high intestinal synthesis, supplementation is not necessary under most circumstances although horses on grass hay and oats diet are probably getting less than the recommended 20 mg/kg of diet. Exercising horses, especially endurance and racing animals, probably require additional riboflavin.

Deficiencies

Horses on normal diets rarely develop life-threatening B vitamin deficiency symtoms. Even if a horse does not eat, the organisms in the intestinal tract supply these vitamins. Symptoms of inadequate amounts of B vitamins are possible. Specifically, inadequate amounts of riboflavin will compromise the horse's ability to generate energy aerobically and may result in subpar performances in endurance and racing horses. At one time it was thought that riboflavin deficiency might be associated with periodic ophthalmia. However, newer research suggests this is not likely.

Toxicity

Toxicity with B vitamins administered in the feed is virtually nonexistent since the kidneys efficiently eliminate any excess. There is no known toxicity with high doses of riboflavin (but they won't help either).

Interactions

B vitamins generally work in conjunction so if you supplement with one B vitamin you should also supplement the others. Riboflavin especially has many interactions—niacin, pyridoxine and folic acid as well as with the amino acid tryptophan and vitamin K. Conversion of riboflavin to its active form will be suppressed in horses with low thyroid function.

Indications

In general, B vitamin supplementation is indicated in any horse not eating normally (or not eating at all), in horses that are heavily stressed for any reason (age, surgery, injury, infection, shipping, etc.), horses with a history of intestinal upset and in horses being heavily exercised. Diets based on grass hay and oats and high energy requirements increase the likelihood riboflavin supplementation is indicated.

SELENIUM

TRACE MINERAL AND ANTIOXIDANT

Probability of Deficiency Per Use	RDA Per Use	Severity of Deficiencies	Toxicity Risk	Complementary Element	Symptoms of Deficiency
	.5mg			Vitamin C Copper	• Muscular cramping
	.5-.75mg				• Degeneration of muscles
	.5-.75mg				• Degeneration of heart muscle
	1mg				• Oxidation of body fat stores
	1mg				• Dummy foals, problems swallowing, nursing, moving about
					• Symptoms suggesting hypothyroidism

Supplementation Recommendation

USE / DIET					
	1mg	1mg	1-2mg	2+mg	2+mg
	1mg	1mg	1-2mg	2+mg	2+mg
	1mg	1mg	1-2mg	2+mg	2+mg
	1mg	1mg	1-2mg	2+mg	2+mg
	1mg	1mg	1-2mg	2+mg	2+mg
	1mg	1mg	1-2mg	2+mg	2+mg

Sources

Selenium is present in all grains and hays in amounts proportional to the selenium content of the soils on which they were grown. The best natural sources are wheat bran, alfalfa hay, dehydrated Brewer's grains, rice bran (with germ), sorghum and soybean meal (44% protein).

Functions

Exposure to drugs, chemicals, preservatives and inhaled impurities generate substances called free radicals. Free radicals are electrically charged molecules that attack normal body tissues to steal an electrical charge that will restore them to a normal, uncharged state. The damaged and now charged cell in turn attacks its neighbors, setting up a chain reaction of cellular damage. The healthy body constantly fights off invasion by bacteria, viruses and other organisms. While trying to protect the body, the immune system can generate free radicals, which are potentially as damaging as any outside threat. Free radicals may be responsible for many uncomfortable symptoms of infection/inflammation. Exercise also results in free radical production that damages the exercising tissue, causing the muscle aching and fatigue. Antioxidants can neutralize free radicals before they damage normal body tissues. Selenium, together with vitamin E, is required for the normal production of the antioxidant enzyme glutathione peroxidase which helps maintain immune system activity and protect muscles from oxidative damage. In laboratory animals, selenium has been found to be necessary for the conversion of thyroid hormone to its active form.

Supplementation

Situations calling for increased intake of antioxidants include infections, wounds, injuries such as sprains and strains, stressful situations such as shipping or change in environment, living in a polluted environment and, especially, exercise. How much selenium to supplement is somewhat controversial. The NRC has stayed with an upper safe limit of 2 ppm (2 mg for each kg of diet consumed) while many nutritionists will tell you they are not concerned about selenium intake until it reaches the 5 ppm mark. The official NRC requirement of 0.1 ppm in the diet is probably adequate for horses at maintenance. However, exercise greatly increases the requirement. There is also growing interest in possible health benefits provided by feeding selenium at higher than the minimal requirements. Further confusing the picture is that the form of selenium in the supplement will influence how well it is absorbed. Sodium selenite is not absorbed as well as selenomethionine. The sodium selenite is a mineral form (the kind found in soil) while selenomethionine is one of the forms that occurs inside the hays and grains. For now, stay within the dosage ranges listed above and use a supplement that contains selenomethionine (or selenocystine) to guarantee good absorption.

Deficiencies

General signs of less than optimal antioxidant intake include poor stress tolerance, exercise related problems including subpar performance, frequent infections, poor wound healing. Selenium deficiency specifically impacts the immune system (infections, poor wound healing) and exercise performance. Selenium, with vitamin E, is widely used to treat and prevent tying-up. This practice is also controversial since research has not been able to explain why selenium might help treat and prevent tying-up. Part of the problem is that tying-up may have more than one cause. Given that most areas of this country have soils borderline to deficient in selenium, use of vitamin E and selenium in horses showing symptoms of muscle problems is certainly a reasonable part of therapy. It is at least theoretically possible that selenium deficiency may be at the root of problems suggesting hypothyroidism in performance horses but associated with normal blood T4 levels. Selenium is required by the enzyme systems in the liver that convert the inactive, T4, form of this hormone to the active, T3 form.

Toxicity

In horses not being exercised, supplementation at a rate of as little as 3.3 mg/kg of diet can produce toxicity. Very large doses produce a syndrome of apparent blindness, colic, staggering, diarrhea and symptoms of pain and distress including elevated pulse, elevated breathing rate and sweating. Long term feeding of lower amounts produces a chronic poisoning associated with loss of hair from the mane and tail and separation of the hoof wall from the foot at the coronary band.

Interactions

Many of the antioxidants complement the actions of one or more other types of antioxidants. Selenium requires adequate amounts of vitamin E to be effective. Vitamin C plays a role in maintaining both in an active form.

Indications

Any horse can probably benefit from supplementation with antioxidants in terms of fewer infections, better wound healing, better stress tolerance, improved ability to detoxify drugs and other chemicals. Horses experiencing exercise-related muscle problems are prime candidates for selenium supplementation.

SODIUM* AND CHLORIDE

MAJOR MINERALS

Probability of Deficiency Per Use	RDA Per Use	Severity of Deficiencies	Toxicity Risk	Complementary Element	Symptoms of Deficiency
	7.5 grams			Chloride Potassium Bicarbonate Water	• Dehydration • Weakness • Mental confusion • Heat exhaustion and collapse
	7.5-11.25 grams				
	11.25-15 grams				
	15-22.5+ grams				
	15-22.5+ grams			*All chart numbers refer to Sodium.	

Supplementation Recommendation

USE DIET					
	salt block	salt block	salt block	salt block	salt block
	salt block	salt block	salt block	salt block	salt block
	salt block	salt block	salt block	salt block	salt block
	salt block	salt block	salt block	salt block	salt block
	salt block	salt block	salt block	salt block	salt block
	salt block	salt block	salt block	salt block	salt block

Sources

Sodium (and chloride) are present in all common feeds but in varying amounts. Grains contain about 0.03% sodium while hays range from as little as 0.02% to 0.1+%.

Functions

Sodium is the major mineral (electrolyte) in the extracellular space—i.e., in the blood and outside of the cells, bathing the tissues. Sodium's major role is to maintain normal hydration—water content—in the body. Sodium is also important to the contraction of muscles and generation of nerve impulses. No specific functions are described for chloride. Chloride is also primarily located in the extracellular fluids. Chloride ions in the blood have a negative electrical charge while sodium's are positive. The sum of all the electrical charges of the major electrolytes (sodium, potassium, chloride and bicarbonate) determines the electrical charge/pH of the blood.

Supplementation

Levels of sodium (and chloride) naturally available in feedstuffs could probably meet maintenance needs in some but not all diets. Precise determinations of sodium requirements at various work levels and with different levels of loss of sodium (and chloride) in sweat have not been determined for horses. Baseline sodium requirement is estimated at about 7.5 grams and it is assumed chloride needs will be met if sodium needs are met. Most horses will take in enough extra sodium chloride (salt) if provided with a salt block. Horses in light to moderate work under mild weather conditions should consume a 3 pound salt block in about a month. Heavier work loads and/or hot temperatures create more sweating and the salt block should be consumed in about two weeks under such stressful conditions. In cold weather, salt and water intake drop and salt blocks will last about six weeks on the average. If the horse does not come close to these general guidelines, he may need to have salt added to the grain ration or water. *Forcing salt intake in this way, however, will do no good if the horse does not also take in extra water. Never add electrolytes to the horse's water unless you are also providing a bucket of plain water.*

Deficiencies

Deficiency of salt (sodium chloride) results in dehydration which causes weakness, trembling, loss of appetite, excessive thirst, rising body temperature and markedly decreased ability to exercise. As the body overheats, blood is shifted away from the muscles and to the skin to allow for extra cooling. Muscles become deprived of maximal oxygen and glycogen stores become depleted more rapidly. Horses on the diuretic Lasix for prevention of lung bleeding during racing also lose increased amounts of sodium.

Toxicity

Oral salt intake even in excessive amounts will not cause any problems (except increased urination), as long as adequate water is available.

Interactions

Water is the most important companion nutrient for salt. Provision of adequate levels of all other major minerals also helps maintain the horses electrolyte status and a normal pH in the blood and tissues.

Indications

All horses should have access to a salt block at all times, along with constant supply of clean, fresh water. Horses that do not take in adequate salt on their own may need to have it added to the grain or a water source.

SULFUR

TRACE MINERAL

Probability of Deficiency Per Use	RDA Per Use	Severity of Deficiencies	Toxicity Risk	Complementary Element	Symptoms of Deficiency
	0			None	None known. See Methionine.
	0				
	0				
	0				
	0				

Supplementation Recommendation

USE / DIET					
	0	0	0	0	0
	0	0	0	0	0
	0	0	0	0	0
	0	0	0	0	0
	0	0	0	0	0
	0	0	0	0	0

Sources

Sulfur is found at levels of less than 0.1 ppm to 0.2 ppm in bran, common hays and grains (alfalfa slightly higher). Molasses contains about twice this amount (0.43 ppm).

Functions

Sulfur plays a role in the production of energy from carbohydrates, blood clotting and in the formation and health of all connective tissues. It is necessary to form the reinforcing bonds between strands of collagen.

Collagen is the base of all connective tissues and the framework for bones, joint cartilage, ligaments, tendons and hooves.

Supplementation

No recommendations are available for the feeding of sulfur in its mineral form. It is believed the horse must obtain most if not all of his sulfur from eating high quality protein sources which supply the sulfur-containing amino acids methionine, cystine and cysteine.

Deficiencies

Deficiencies of sulfur/sulfur containing amino acids could have many consequences including weakness/abnormalities of tendons, ligaments, bones and joint cartilage, hooves and poor carbohydrate metabolism.

Toxicity

No naturally occurring toxicity has been described, but horses accidentally fed 200 to 400 grams of inorganic (mineral) sulfur developed lethargy and colic within 12 hours, followed by liver failure and lung problems. Two of 12 affected horses died.

Interactions

Excessive sulfur can suppress copper absorption in other species.

Indications

None, for inorganic (mineral form) sulfur. See methionine.

THIAMINE

B VITAMIN

Probability of Deficiency Per Use	RDA Per Use	Severity of Deficiencies	Toxicity Risk	Complementary Element	Symptoms of Deficiency
	0-50mg			Magnesium Other B-vitamins	• Elevated blood lactate and pyruvate
	0-50mg				• Muscle cramping and pain
	50-100mg				• Overexcitability
	100-250mg				• Slow growth
	100-250mg				• Decreased appetite
					• Incoordination

Supplementation Recommendation

USE / DIET					
	0	0	100 mg	250 mg	250 mg
	0	0	50 mg	100 mg	100 mg
	0	0	100 mg	250 mg	250 mg
	0	0	50 mg	100 mg	100 mg
	0	0	100 mg	250 mg	250 mg
	0	0	50 mg	100 mg	100 mg

Sources

A horse's diet derives B vitamins from grains, brans and yeast. Also, the abundant microorganism population in the horse's intestinal tract can manufacture B vitamins, and a horse may absorb a certain amount from this source.

Functions

B vitamins are important to every organ and cell of the horse's body. Thiamine is essential to the proper metabolism of carbohydrates, fats and proteins. Without adequate thiamine, cells are unable to properly process lactate and pyruvate, leading to high levels of these substances in the blood. Thiamine is also needed for cells to copy their DNA and reproduce themselves. Thiamine also functions in the transmission of impulses along nerves.

Supplementation

Because B vitamins are water soluble, they float freely through the fluids of the body and are not stored in any body tissues. Once these vitamins are absorbed into the blood, they circulate and are taken in by cells that need them or eliminated in the urine. Because of this rapid elimination, horses need B vitamins on a daily basis. The NRC officially recommends a daily intake of 2.5 mg/kg of feed per day for inactive horses and 5 mg/kg of feed per day for horses in work. The average diet contains thiamine at a level of about 2.9 mg/kg of feed, maximum. Horses receiving little or no grain will get considerably less. Exercise physiology studies in other species have shown that thiamine needs in horses on high carbohydrate diets and that exercise heavily are much higher than those officially recommended. Our recommendations take into consideration both the likelihood of increased needs with exercise and the fact that horses in work are on higher concentrated carbohydrate diets (i.e., grain) than horses at rest.

Deficiencies

Developing full blown, life-threatening Vitamin B deficiency symptoms is unlikely in horses on normal diets. Even if a horse is not eating, organisms in the intestinal tract are capable of manufacturing these vitamins and some will be absorbed by the horse. However, the horse can develop symptoms of inadequate amounts of B vitamins. Specifically, symptoms of thiamine deficiency include nervousness, irritability, excitability, elevated blood lactate and pyruvate and muscle cramping and pain—possibly even obvious tying-up. Symptoms reported in other species that may apply to the horse also include picky appetite and swelling of the extremities (stocking up).

Toxicity

Toxicity with B vitamins administered in the feed is virtually nonexistent since the kidneys will efficiently eliminate any excess. Thiamine has a calming effect on some horses at higher dosages (500+ mg/day).

Interactions

B vitamins work together, so it is usually advised that if you supplement with one B vitamin you should also supplement the others. If the horse is on a high carbohydrate/grain diet, supplementation with all the B vitamins is advisable. Magnesium is necessary to convert dietary thiamine into its active form inside the body. When being used for its calming effects, this may be enhanced by the addition of supplemental calcium.

Indications

In general, B vitamin supplementation is called for in any horse that is not eating normally (or not eating at all), in horses that are heavily stressed for any reason (age, surgery, injury, infection, shipping, etc.), in horses with digestive problems (upsets the population of organisms in the gut) and in horses being heavily exercised. High grain intake also calls for supplemental thiamine. Thiamine is often used in higher doses (500+ mg per day) as part of the preventative treatment of tying-up. The same dosage level (500 to 1,000 mg/day) is often tried, with or without supplemental calcium, to calm horses that are nervous, easily distracted, jumpy and difficult to work around.

VITAMIN A

ANTIOXIDANT AND FAT SOLUBLE VITAMIN

Probability of Deficiency Per Use	RDA Per Use	Severity of Deficiencies	Toxicity Risk	Complementary Element	Symptoms of Deficiency
	15,000 to 30,000 IU			Vitamin E Zinc Fat Protein	• Poor night vision • Decreased fertility • Thickening of skin and cornea • Frequent respiratory infections • Weakness of bones • Slow growth
	15,000 to 30,000 IU				
	15,000 to 30,000 IU				
	15,000 to 30,000 IU				
	15,000 to 30,000 IU				

Supplementation Recommendation

USE / DIET					
	30,000 IU	30,000 IU	30,000 IU	30,000 IU	30,000 IU
	30,000 IU	30,000 IU	30,000 IU	30,000 IU	30,000 IU
	0 to 10,000	0 to 10,000	0 to 10,000	0 to 10,000	0 to 10,000
	0 to 10,000	0 to 10,000	0 to 10,000	0 to 10,000	0 to 10,000
	10,000 to 20,000 IU	10,000 to 20,000 IU	10,000 to 20,000 IU	10,000 to 20,000 IU	10,000 to 20,000 IU
	10,000 to 20,000 IU	10,000 to 20,000 IU	10,000 to 20,000 IU	10,000 to 20,000 IU	10,000 to 20,000 IU

Sources

The major source of vitamin A in the diet is fresh pastures and top quality alfalfa hay. Grass hays contain much more vitamin A than grains but not enough to maintain normal vitamin A blood levels over the winter in unsupplemented horses. Vitamin A in natural foods is present primarily as carotene—a chemical that is two vitamin A molecules hooked together. This is broken apart into vitamin A in the intestinal tract and possibly elsewhere in the body.

Functions

Vitamin A is important to the health of the body's mucus membranes—the lining of the respiratory, digestive and reproductive tracts. Its importance in horses is not as great as in other species where cholesterol abnormalities are common and may result in heart disease. However, vitamin A has other critical roles. It is necessary for the production of sperm and eggs. It is absolutely essential to eye health, and one of the earliest symptoms of deficiency is difficulty seeing at night. Vitamin A also is needed for normal bone growth and healthy skin and hair.

Supplementation

Vitamin A has been well studied in horses, although with varying results, and there is some confusion about what is a safe dose. The current NRC recommendations are that all horses need 30 to 60 IU/kg of body weight of vitamin A (or its equivalent in natural carotenes). Horses maintained on pasture and high quality alfalfa hay probably require little or no additional vitamin A but may experience a drop in blood levels (and with that an increased risk of respiratory infections) in the winter months. Because vitamin A is so well studied, complete feeds are well supplied with this vitamin, as are most supplements made for horses. Horses on a grass hay based or mixed hay diet will likely benefit from vitamin A supplementation. Horses that are stabled year round and have no access to fresh grass are in the highest risk group for inadequate vitamin A intake.

Deficiencies

Symptoms of deficiency include poor night vision, dry and thickened skin, poor hair coat and/or hair loss, frequent respiratory infections, abnormal development of bone in young animals and weakening of bones with chronic deficiency in older animals. Vitamin A and carotene have received considerable attention because of their importance to normal reproduction. The blood and tissue levels of vitamin A are likely to be at their lowest at those times of the year when breeders of Thoroughbreds, Standardbreds and racing quarter horses would like to begin breeding—i.e., prior to growth of fresh pastures. Nature would like to have horses breed in late spring and early summer but man wants bigger yearlings! Short term, high dose supplementation (less than 6 months) of breeding stallions and mares prior to and during early breeding seasons appears to increase fertility.

Toxicity

Excess vitamin A intake can cause fragile bones, abnormal deposits of bone anywhere in the body, loss of patches of skin, rough hair coat, weakness and depression. In an experiment using ponies, unthriftiness/doing poorly was noted after only 15 weeks on excess vitamin A. The NRC proposes an upper safe limit of 80,000 IU/day for a 500 kg horse at maintenance for long term (over 6 months) supplementation. This refers only to vitamin A. Intake of large amounts of carotene rich alfalfa hay has never been noted to cause toxicity. The horse's body probably prevents toxicity by modifying the enzyme that converts carotene to active vitamin A. Toxicity is a real possibility if you use multiple supplements as many contain large amounts of vitamin A (read the labels), especially on an alfalfa/pasture based ration. Decrease the possibility of toxicity by using carotene supplements or natural carotene sources instead of vitamin A. One pound of carrots per day supplies 30,000 IU of vitamin A.

Interactions

Many of the antioxidants complement the actions of one or more other types of antioxidants. Vitamin E protects vitamin A. Zinc as well as adequate dietary fat and protein are needed for vitamin A to be converted to its active form.

Indications

Vitamin A is very important for normal skeletal development in growing horses and remodeling of bone in response to exercise in older horses. Horses prone to skin problems may benefit from vitamin A supplementation. Short term high dose vitamin A/carotene supplementation is often recommended for breeding stock.

VITAMIN C

ANTIOXIDANT

Probability of Deficiency Per Use	RDA Per Use	Severity of Deficiencies	Toxicity Risk	Complementary Element	Symptoms of Deficiency
	 0			Vitamin E Bioflavinoids Manganese Copper Lipoic acid	• Weak tendons and ligaments • Joint pain/arthritis • Frequent infections and impaired ability to fight infections • Gum disease • Allergies • Bleeding from the lungs with exercise • Abnormal development of bone and joints, fragile bone
	 0				
	 0				
	 0				
	 0				

Supplementation Recommendation

USE / DIET					
	4.5 grams	4.5	7.0	7 to 10	7 to 10
	4.5 grams	4.5	7.0	7 to 10	7 to 10
	4.5 grams	4.5	7.0	7 to 10	7 to 10
	4.5 grams	4.5	7.0	7 to 10	7 to 10
	4.5 grams	4.5	7.0	7 to 10	7 to 10
	4.5 grams	4.5	7.0	7 to 10	7 to 10

Sources

Very little information is available on the levels of vitamin C in equine diets although fresh grasses are probably the best, if not the only significant, source. However, there are no known cases of full-blown vitamin C deficiency (caused by an absolute absence of vitamin C in the diet) in horses. It is presumed that horses, like many other species, can manufacture their own supply of vitamin C, although the extent to which they can do this is unknown.

Functions

Free radicals are electrically imbalanced molecules that attack normal body tissues to steal an electrical charge that will restore them to a normal state. The damaged cell in turn attacks its neighbors in the same way, setting up a chain reaction of cellular damage. Free radicals are generated as a waste product when the immune system defends the body from invasion by bacteria, viruses and other organisms. Free radicals may be responsible for many uncomfortable symptoms of infection/inflammation (pain, swelling) and for other symptoms of viral infections. Exercise also results in free radical production that damages the exercising tissue and may cause muscle aching and fatigue. Antioxidants neutralize free radicals before they can damage normal body tissues. Vitamin C is the major antioxidant involved in protecting the tissues of the respiratory system from infections and damage by chemicals. Vitamin C is also a major contributor to the horse's ability to fight off infections. In addition, vitamin C is required for the production of normal connective tissues, the thin sheets of tissue that surround muscle bellies and tendons. Connective tissue is found in the walls of blood vessels and forms the framework for bones, tendons, ligaments and joint capsules. Vitamin C is also required for the production of epinephrine—the chemical released in high amount with stress or fear (and exercise)—and the steroid hormones such as cortisol.

Supplementation

Because vitamin C can be manufactured by the horse and horses do not develop full-blown deficiency states, vitamin C has received very little attention from nutritionists. We do know that horses which are kept stabled (denied access to pasture) and horses being heavily exercised can experience drops in blood level of vitamin C to virtually zero. The list of problems associated with inadequate vitamin C levels, including susceptibility to viral infections, tendon and ligament problems, joint problems, breathing problems and bleeding from the lungs with strenuous exercise reads like a list of many of the most common health problems of horses. These observations make a very strong circumstantial case for supplementation with vitamin C. The 4.5 gram daily dose minimum is recommended because it has been shown that is the minimum required to make any difference in vitamin C blood levels. Blood levels of a vitamin will not begin to rise until the tissue needs for it have been met.

Deficiencies

General signs of suboptimal antioxidant intake include poor stress tolerance, exercise related problems including subpar performance, frequent infections, poor wound healing. Over the long term, suboptimal antioxidant intake is believed to contribute to premature aging, cancer and health problems such as heart disease and diseases of the blood vessels. As mentioned above, the consequences of inadequate vitamin C include many of the most common health problems of horses.

Toxicity

With the exception of vitamin A, antioxidants in very large doses are basically nontoxic. Vitamin C in very large doses has been reported to cause digestive upset in people, but horses tolerate vitamin C in the dosage ranges listed above very well. If you have a sensitive individual horse, vitamin C is available in a buffered form, which helps prevent upset. Extremely high intake of vitamin C (15 to 20 grams a day for a horse) can interfere with absorption of vitamin B12 from the diet. However, because the horse does not rely on diet to provide his B12, this is much less likely.

Interactions

Many antioxidants complement the actions of one or more other types of antioxidants. Among vitamin C's roles is to reactivate the antioxidants vitamin A and vitamin E after they have neutralized a free radical. The bioflavinoids have a similar complementary role with vitamin C. For proper connective tissue, joint, bone and blood vessel health, vitamin C works with other nutrients as well, including the minerals copper, manganese and zinc. The nutrient lipoic acid can also recycle vitamin C. Conversely, iron in the diet may neutralize vitamin C before it can be absorbed.

Indications

Any horse can probably benefit from supplementation with antioxidants in terms of fewer infections, better wound healing, better stress tolerance, improved ability to detoxify drugs and other chemicals and possibly even a longer life. Although controlled research has not been done on horses, many racing horses with the problem of lung bleeding have benefited from vitamin C and bioflavinoids. Regular users of antioxidants, including vitamin C, report far fewer respiratory problems/infections. Vitamin C is also indicated for the treatment and prevention of bone, joint, tendon and ligament problems.

VITAMIN D

FAT SOLUBLE VITAMIN

Probability of Deficiency Per Use	RDA Per Use	Severity of Deficiencies	Toxicity Risk	Complementary Element	Symptoms of Deficiency
	0			Calcium Phosphorus	• Loss of appetite • Weakness • Abnormal growth and abnormal bone development
	0				
	0				
	0				
	0				

Supplementation Recommendation

USE / DIET					
	0-5,000 IU	0-5,000 IU	0-5,000 IU	0-5,000 IU	0-5,000 IU
	0-5,000 IU	0-5,000 IU	0-5,000 IU	0-5,000 IU	0-5,000 IU
	0-5,000 IU	0-5,000 IU	0-5,000 IU	0-5,000 IU	0-5,000 IU
	0-5,000 IU	0-5,000 IU	0-5,000 IU	0-5,000 IU	0-5,000 IU
	0-5,000 IU	0-5,000 IU	0-5,000 IU	0-5,000 IU	0-5,000 IU
	0-5,000 IU	0-5,000 IU	0-5,000 IU	0-5,000 IU	0-5,000 IU

Sources

Vitamin D is not obtained directly from the diet. Precursors of the vitamin are found in plants and also manufactured in the horse's tissues. These precursors undergo a further transformation when they are present in the skin and the skin is exposed to sunlight. This second compound, formed after exposure to sunlight, is further metabolized in the liver and kidney to the active form of the hormone.

Functions

Vitamin D controls the absorption of calcium from the intestine, movement of calcium in and out of bone and the amount of calcium that is kept or excreted in the kidney.

Supplementation

Because full-blown vitamin D deficiency states do not exist in horses under natural conditions, very little study has been directed at this vitamin. Like vitamin C and vitamin K, it is presumed the horse can make enough on his own. The likelihood that horses with regular exposure to as little as one hour of sunlight per day will need supplemental vitamin D is quite low. The horse also stores vitamin D in the liver and kidney, for use when supplies become low. At greatest risk of borderline to inadequate vitamin D levels are foals that are born quite early in the year and must be kept indoors for prolonged periods. These foals do not have the benefit of liver and kidney stores that older animals have. Furthermore, their rapid growth means they have a much greater requirement for adequate vitamin D than mature animals. Studies on pony foals maintained indoors and deprived of vitamin D in their diets did indeed confirm symptoms of deficiency. Loss of appetite occurred first, followed by unwillingness/inability to stand. Examination of the bones after the ponies were euthanized confirmed they were abnormal. Symptoms could be prevented by feeding 1,000 IU of vitamin D/kg of dry diet to the very young ponies or 500 IU/kg of dry diet to older pony foals. That 500 IU/kg of dry diet number was the one used in making supplementation recommendations above. Since horses are known to have lower blood levels of vitamin D than many other animals and because supplementation even in apparently normal animals resulted in improved calcium absorption, supplementation, especially of young animals, may be beneficial. Horses with fractures may also benefit from supplementation with vitamin D. Vitamin D improves fracture healing in apparently normal animals of other species.

Deficiencies

Artificially induced severe deficiencies result in abnormal bone growth, loss of appetite and inability to stand.

Toxicity

Vitamin D toxicity is a very real possibility. Excessive vitamin D causes calcium to be deposited in the blood vessels, heart and other tissues such as skin. Calcium is pulled out of bones, making them weak, extensive kidney damage occurs and young ponies on an intake of 3,300 IU/kg of diet died within four months. Other studies have had similar results and confirmed the toxicity of vitamin D. The NRC has proposed an upper safe limit for supplementation of 2,200 IU/kg of dry diet which would be 11,000 IU/day for a 500 kg horse at maintenance to 22,000 units/day to the same horse in heavy work.

Interactions

Vitamin D is the master vitamin controlling calcium metabolism.

Indications

Judicious supplementation may be of benefit to growing animals and to animals with fractures. Horses in heavy work, undergoing bone remodeling, may also benefit. Vitamin D supplementation is indicated for any horses that must be kept indoors for prolonged periods (e.g., greater than 4 weeks).

VITAMIN E

ANTIOXIDANT AND FAT SOLUBLE VITAMIN

Probability of Deficiency Per Use	RDA Per Use	Severity of Deficiencies	Toxicity Risk	Complementary Element	Symptoms of Deficiency
	0			Selenium Vitamin C Zinc	• Impaired immunity • Excessive inflammatory reaction to injuries or infections • Muscle weakness, cramping or tying-up
	0				
	800-1,000 IU				
	800-1,000 IU				
	800-1,000 IU				

Supplementation Recommendation

USE / DIET					
🌾	500 I.U.	1,000 I.U.	2,000 I.U.	2,000-5,000	2,000-5,000
🌾 OATS	500 I.U.	1,000 I.U.	2,000 I.U.	2,000-5,000	2,000-5,000
🍀	500 I.U.	1,000 I.U.	2,000 I.U.	2,000-5,000	2,000-5,000
🍀 OATS	500 I.U.	1,000 I.U.	2,000 I.U.	2,000-5,000	2,000-5,000
🌾🍀	500 I.U.	1,000 I.U.	2,000 I.U.	2,000-5,000	2,000-5,000
🌾🍀 OATS	500 I.U.	1,000 I.U.	2,000 I.U.	2,000-5,000	2,000-5,000

Sources

High natural sources of vitamin E include vegetable oils. However, oils must be raw/unprocessed or cold processed to retain their vitamin E content. Oils purchased from a grocery store shelf have been stabilized to make them suitable for cooking, destroying the vitamin E. (Extra virgin olive oil is the exception to this—extra virgin by definition means cold processed.)

All grains, hays and grasses contain vitamin E. However, vitamin E activity decreases with time of storage and exposure to too high a moisture content. In addition, only whole grains retain much vitamin E activity. Crimping, rolling, crushing, etc., greatly accelerates breakdown of vitamin E.

Functions

Vitamin E is an antioxidant, which neutralizes free radicals before they can damage normal body tissues. Also a fat soluble vitamin, it is attracted to areas that contain fat, such as the walls of cells and membranes surrounding the important structures inside cells. Vitamin E protects and stabilizes every membrane in the body and protects muscle cells from the damaging free radicals generated during exercise. The more the horse exercises, the more vitamin E he needs. In the immune system, vitamin E protects white cells from the damaging effects of the destructive enzymes used to dissolve dead tissues and kill invading microorganisms (bacteria, viruses, etc.). If there is inadequate vitamin E, these potent chemicals will cause as much damage to the normal tissues in the area as they do to dead or injured tissue and invaders.

Supplementation

Antioxidants can be helpful in dealing with infections, wounds, injuries such as sprains and strains, stressful situations such as shipping, living in a polluted environment and, especially, exercise. Vitamin E is one vitamin even the NRC recognizes should be supplemented. Blood vitamin E levels drop off sharply during those times of the year when horses do not have access to good grass at pasture. NRC recommendations now are for supplementation of pregnant, growing, lactating and working horses at levels which have been shown to be required for normal function of the immune system. However, they do not take into account the increased antioxidant needs of horses that are exercising. The above recommendations for exercising horses take that need into consideration. There are studies that report no benefit from exercising horses with supplemental vitamin E, but the experience of vets and trainers working with these horses indicate otherwise. It will not improve performance per se (e.g., no greater speed), but is needed to protect the cells from damaging free radicals generated when working at speed.

Deficiencies

Less than optimal antioxidant intake may result in poor stress tolerance, exercise related problems including subpar performance, frequent infections and poor wound healing. Inadequate vitamin E has been proven to result in depressed immune function in horses. Supplementation of 1,000 IU/day has been effective in preventing muscle damage caused by the chasing and capture of wild horses. Dosages much higher (5,000 to 10,000 IU) are used by some trainers and veterinarians to treat exercise related muscle pain and to treat and prevent tying-up.

Toxicity

With the exception of vitamin A, antioxidants even in very large doses are basically nontoxic. Toxicity of even high doses of vitamin E has never been reported for healthy animals of any species. However, because vitamin E is a fat soluble vitamin stored in the liver (although for apparently a very short time), and theoretically at least could interfere with other fat soluble vitamins, the NRC has proposed an upper safe limit of 10,000 IU/day.

Interactions

Vitamin C can restore vitamin E to an active state after it has "captured" a free radical. For vitamin E to work well there must be adequate levels of the trace mineral selenium. Zinc, another trace mineral with antioxidant properties, also enhances the activity of vitamin E. Conversely, the presence of iron, a highly reactive substance, in the same feed as vitamin E will neutralize the vitamin E. Feeding polyunsaturated fats and oils increases the requirements for vitamin E. The vegetable oils commonly fed to horses contain over 80% polyunsaturated fats on the average (flax/linseed and canola oils have more, palm and coconut oils much less). Horses on high fat diets (especially with muscle problems) require increased vitamin E.

Indications

To be safe, even horses at maintenance should receive a minimum of 500 IU/day of vitamin E. Appropriate dosage will increase with the magnitude of work the horse must do. It is very important to supplement selenium at the same time as vitamin E. A ratio of 1 mg of selenium for each 1,000 IU of vitamin E works well in most cases. If horses reside in a known high selenium area, however, check with the veterinarian before supplying any supplemental selenium. Because of vitamin C's function as a recycler of vitamin E, supplement vitamin C at the same time. A ratio of 1 gram of vitamin C for each 1,000 mg of vitamin E seems to work well.

VITAMIN K

FAT SOLUBLE VITAMIN

Probability of Deficiency Per Use	RDA Per Use	Severity of Deficiencies	Toxicity Risk	Complementary Element	Symptoms of Deficiency
	0			None	• Never described in healthy horses
	0				• Clotting abnormalities in other species
	0				
	0				
	0				

Supplementation Recommendation

USE / DIET					
	0	0	0	0	0
	0	0	0	0	0
	0	0	0	0	0
	0	0	0	0	0
	0	0	0	0	0
	0	0	0	0	0

Sources

Vitamin K precursors (phylloquinones) are in abundant supply in plants. In addition, the bacteria in the intestinal tract can process these precursors further to a form called menaquinones or menandiones. Precursors are further chemically altered in the horse's liver to produce the active form of vitamin K, which is essential to normal blood clotting.

Functions

As noted, vitamin K is essential to normal blood clotting, although vitamin K responsive bleeding problems can result in horses that are given anticoagulants such as warfarin to treat laminitis. Moldy sweet clover hay also produces a toxic chemical that causes the same problem (which is why clover hays are not recommended for horses).

Supplementation

Although vitamin K is commonly found in equine supplements, there is no NRC established required dietary level. There are also no described cases of vitamin K dependent clotting abnormalities in horses with normal livers.

Deficiencies

Deficiencies of vitamin K probably do not occur in normal horses. Vitamin K has been tried in attempts to decrease the problem of bleeding from the lungs with exercise but has not been shown to be effective. However, it is at least theoretically possible that horses with chronic digestive tract upsets and altered intestinal tract microorganisms could have inadequate absorption of vitamin K precursors. However, the natural diet is the horse's other abundant source and vitamin K in excess of amounts needed is well stored in the liver. In people excessive calcium intake, with calcium:phosphorus ratios of higher than 2:1 appear to interfere either with absorption of precursors or synthesis in the intestine and can cause bleeding in the intestinal tract. This has not been recognized in horses. If it does occur, horses on straight alfalfa hay would be at risk. In people high intakes of vitamin E can interfere with vitamin K function and absorption. The equine equivalent would be over 12,000 IU/day.

Toxicity

Oral administration of phylloquinones, vitamin K precursors, appears to be basically harmless. The use of the menandione form in people is potentially toxic, causing anemia and the destruction of red blood cells. However, this does not appear to be much of a concern in horses. What is dangerous is the use of injectable menandione in horses, either intravenously or intramuscularly. Injection of this compound, even at doses recommended by the manufacturer, has caused severe kidney damage. Bottom line is that feeding vitamin K appears to be safe but is likely to be unnecessary and ineffective.

Interactions

Vitamin K does not have any recognized helper nutrients.

Indications

At present, there are no indications for supplementation with vitamin K in the horse.

ZINC

TRACE MINERAL AND ANTIOXIDANT

Probability of Deficiency Per Use	RDA Per Use	Severity of Deficiencies	Toxicity Risk	Complementary Element	Symptoms of Deficiency
	250 mg			Copper	• Poor wound healing
	250 mg				• Susceptibility to infections
	250-375 mg				• More severe symptoms with infections
	375-500 mg				• Impaired carbohydrate metabolism with decreased performance
	375-500 mg				• Abnormal skin and hooves
					• Bone and joint problems
					• Abnormal sensation
					• Abnormal thyroid hormone metabolism

Supplementation Recommendation

USE / DIET					
	100 mg	100 mg	100-275	275-500	275-500
	100 mg	100 mg	100-275	275-500	275-500
	100 mg	100 mg	100-275	275-500	275-500
	100 mg	100 mg	100-275	275-500	275-500
	100 mg	100 mg	100-275	275-500	275-500
	100 mg	100 mg	100-275	275-500	275-500

Sources

Zinc is present in all grains and hays but at levels insufficient to meet even the NRC's recommendations. Of the commonly used feeds, only wheat bran supplies enough zinc. Commercial feeds are usually well supplemented, however.

Functions

Free radicals are electrically imbalanced molecules that attack normal body tissues to steal an electrical charge that will restore them to a normal neutral state. The damaged cell in turn attacks its neighbors in the same way, setting up a chain reaction of cellular damage. These potentially damaging substances may be created when the immune system fights bacteria, viruses and other ogranisms. Free radicals may cause

Functions — continued

the uncomfortable symptoms of infection/inflammation (pain, swelling) and the sore throats, runny noses and other symptoms of viral infections. Exercise may also cause free radical production that damages the exercising tissue. It may be a significant cause of the muscle aching and fatigue that follows heavy exercise even in a normal animal. Antioxidants are substances present in the diet and manufactured by the body that can neutralize free radicals before they can damage normal body tissues. Zinc functions as part of the superoxidase dismutase antioxidant enzyme system. Zinc is important to the maintenance of the enzyme system in the liver which converts thyroid hormone into its active form and could account for performance horses who have symptoms that suggest hypothyroidism but normal levels of T4 (the inactive form of thyroid hormone) in their blood. Zinc is extremely important to immune function in both fighting invading organisms and the control of symptoms related to inflammation. Health of the skin and feet depends greatly on zinc. Zinc is also involved in successfully metabolizing carbohydrates and the normal functioning of the sensory nerves of the body, as well as in bone and joint health.

Supplementation

Antioxidants are a perfect example of how minimal intake of a nutrient may not be fatal but matching intake to needs results in greatly improved health. Situations which call for increased intake of antioxidants include infections, wounds, injuries such as sprains and strains, stressful situations such as shipping or change in environment, living in a "polluted" environment and, especially, exercise. Zinc's multiple functions make it critical to maintain adequate levels of this mineral. Zinc deficiency is probably one of the most widespread and most overlooked health problems of horses. Normal diets supply only about 3/5 of the minimum recommended by the NRC. Although the research has not been done in horses, studies with laboratory animals and humans confirm that trace mineral requirements increase with exercise. Hence the higher levels for performance horses.

Deficiencies

Inadequate antioxidant intake may result in poor stress tolerance, exercise related problems including subpar performance, frequent infections, poor wound healing. Over the long term, suboptimal antioxidant intake is believed to contribute to premature aging, cancer and health problems such as heart disease and diseases of the blood vessels. With specific reference to zinc, poor wound healing, decreased fertility, increased susceptibility to infections, symptoms suggesting hypothyroidism and suboptimal performance (related to defective carbohydrate metabolism) could be seen.

Toxicity

Zinc supplementation has very low risk of toxicity. It is estimated that amounts up to 1,000 ppm (mg/kg) which would be 5,000 mg for a horse at maintenance, will be tolerated without problem. Also, supplement copper to a ratio of 3:1 zinc: copper to avoid competition for absorption.

Interactions

Many antioxidants complement the actions of one or more other types of antioxidants. Zinc functions with copper in the superoxide dismutase antioxidant system. Both zinc and selenium are needed for proper conversion of thyroid hormone to its active form. Zinc may interfere with the absorption of copper when both are in their mineral forms and a ratio of zinc:copper of 3:1 has been suggested to maximize proper absorption. Use of copper and zinc chelated (bound) to a protein molecule eliminates absorption problems–e.g., zinc methionine.

Indications

Any horse can probably benefit from supplementation with antioxidants in terms of fewer infections, better wound healing, better stress tolerance, improved ability to detoxify drugs and other chemicals and possibly even a longer life. The varied and important other functions of zinc, plus the fact that virtually any commonly fed diet (except supplemented feeds) is likely to be deficient, makes zinc supplementation highly advisable.

CHAPTER 4

A TO Z OTHER

Hunters and jumpers usually fall in the moderate use category.

ABOUT THIS CHAPTER

There are many supplements on the market, and mentioned in this book, which do not appear in the Nutrient Requirements for Horses, published by the federal government, and therefore do not have an established daily requirement. True deficiency states may not exist for many of these supplements. In other words, the horse can live without them but might benefit from their use under specific circumstances.

Some of these nutrients are actual proteins or minerals that appear in the basic diet but have not had their requirement established. Others are specific metabolites—chemicals, enzymes, proteins that appear and are used in the body during the course of a reaction such as the burning of fuels for energy production or the building of muscle tissues and joints. Still others are natural sources of vitamins, minerals and proteins that are not commonly found in the horse's diet.

This section follows a similar format to the A to Z of vitamins and minerals. It will give you the name of the nutrient, probability of benefit for horses at specific activities, toxicity (if any), interactions with other nutrients, specific conditions that might respond to the supplement and a full description of what it is and how it is believed to work.

LEGEND

Use Icons	Probability of Deficiency					Severity of Deficiency & Risk of Toxicity
Maintenance	None	Little	Moderate	High	Heavy	None
Light	None	Little	Moderate	High	Heavy	Little
Moderate	None	Little	Moderate	High	Heavy	Moderate
High	None	Little	Moderate	High	Heavy	Serious
Heavy	None	Little	Moderate	High	Heavy	

BETA-HYDROXY-BETA-METHYLBUTYRATE

PROBABILITY OF BENEFIT	DOSAGE	OVERDOSE RISK	COMPLEMENTARY NUTRIENTS	INDICATED FOR
	16 to 40 gm/day		Vitamin C	• Infections • Lung bleeding • Allergies • Heavy exercise • Antioxidant protection

Sources

Small amounts of HMB are found in foods but most is derived from the metabolism of the branched chain amino acid leucine, found in the protein of normal foods or given as a supplement.

Functions

HMB is used in the body to make cholesterol. Although cholesterol has negative health connotations for most people, it is critical to the formation of hormones necessary for the horse to adapt to intense exercise. Feeding HMB may also encourage the metabolic pathways involving branched chain amino acids and energy generation with exercise. Such supplements result in less muscle loss with exercise, improved muscle mass in response to exercise, more stable blood glucose and less lactate production.

Supplementation

Supplementation is recommended at a rate of 5 gm twice a day.

Deficiencies

No deficiency state per se has been described. However, unsupplemented compared to supplemented horses show a drop in blood cholesterol with training, increased lactate production compared to controls and less muscle mass.

Toxicity

None known at suggested doses. Do not use concurrently with branched chain amino acids.

Interactions

HMB will complement other aerobic energy enhancers such as carnitine and DMG. Adequate vitamin B6 is needed for proper metabolism of all proteins.

Indications

HMB is indicated for all horses in heavy work, especially in the training stages. It should prove most useful for horses with high aerobic energy needs such as endurance, 3-day event and Thoroughbred race horses.

BIOFLAVINOIDS

PROBABILITY OF BENEFIT	DOSAGE	OVERDOSE RISK	COMPLEMENTARY NUTRIENTS	INDICATED FOR
	16 to 40 gm/day		Vitamin C	• Infections • Lung bleeding • Allergies • Heavy exercise • Antioxidant protection

Sources

Bioflavinoids are present in a wide variety of plants, although it is most plentiful in those that contain significant amounts of vitamin C.

Functions

Exposure to drugs, chemicals, preservatives and inhaled impurities in the air result in the generation of substances called free radicals. Free radicals are electrically imbalanced molecules that attack normal body tissues to steal an electrical charge that will restore them to a normal state. The damaged cell in turn attacks its neighbors in the same way, setting up a chain reaction of cellular damage. The healthy body is constantly fighting off invasion by a large variety of bacteria, viruses and other organisms. When the immune system carries out this function, a normal waste product is the generation of free radicals which are potentially as damaging as any outside threat. Free radicals may be responsible for many of the very familiar and uncomfortable symptoms of infection/inflammation (pain, swelling) and for symptoms of viral infections. Exercise and the generation of energy also result in free radical production that damages the exercising tissue. This process is considered to be a significant cause of the muscle aching and fatigue that follows heavy exercise, even in a normal animal. Antioxidants are substances present in the diet and manufactured by the body whose function is to neutralize free radicals before they can damage normal body tissues. If the production of free radicals exceeds the supply of antioxidants, tissue damage will result. Bioflavinoids are known to enhance the activity of vitamin C, especially with respect to strengthening blood vessel walls, and are believed to have the ability to neutralize free radicals. There are many different bioflavinoids, all shown in experiments to have potent antioxidant and anti-inflammatory activities. The bioflavinoids rutin and hesperidin in particular have been reported helpful in controlling exercise induced lung bleeding in horses. Catechin has mild anti-inflammatory effects and is helpful in decreasing the release of histamine in allergic situations. Quercitin is of interest also in inflammatory and allergic diseases as it stabilizes mast cells (which contain histamine) and inactivates many inflammatory enzymes. It may also interfere with viruses' ability to reproduce.

Supplementation

Antioxidants are a perfect example of how minimal intake of a nutrient may not be fatal but matching intake to needs results in greatly improved health. Situations which call for increased intake of antioxidants include infections, wounds, injuries such as sprains and strains, stressful situations such as shipping or change in environment, living in a polluted environment and, especially, exercise. There is little information or for-

Supplementation

mal research available concerning bioflavinoid use in the horse. The numbers above are based on ratios of bioflavinoid to vitamin C in human supplements and supplements believed to be beneficial for horses.

Deficiencies

General signs of suboptimal antioxidant intake include poor stress tolerance, exercise related problems including subpar performance, frequent infections, poor wound healing. Over the long term, suboptimal antioxidant intake is believed to contribute to premature aging, cancer and health problems such as heart disease and diseases of the blood vessels. Bioflavinoids have not been shown to be essential to life—meaning a diet which does not contain any will not result in any recognizable deficiency state. In general, supplementation with bioflavinoids can be expected to prevent the same problems associated with inadequate vitamin C (see Vitamin C).

Toxicity

With the exception of vitamin A, antioxidants even in very large doses are basically nontoxic. There are no known toxicities associated with the use of naturally occurring mixtures of bioflavinoids (i.e., citrus extracts).

Interactions

Many of the antioxidants complement the actions of one or more other types of antioxidants. Bioflavinoids support the antioxidant status of the body and in particular enhance the activity of vitamin C.

Indications

Any horse can probably benefit from supplementation with antioxidants in terms of fewer infections, better wound healing, better stress tolerance, improved ability to detoxify drugs and other chemicals and possibly even a longer life. Indications for bioflavinoid supplementation are essentially the same as those for vitamin C.

CHONDROITIN SULFATE

PROBABILITY OF BENEFIT	DOSAGE	OVERDOSE RISK	COMPLEMENTARY NUTRIENTS	INDICATED FOR
	7.5 gm/day		Glucosamine Vitamin C MSM Manganese Copper Other GAGs S-adenosylmethionine	• Arthritis • Tendon and ligament problems • Old age

Sources

Not present in normal equine diet. Chondroitin sulfate is obtained by processing animal products high in connective tissue/cartilage such as bovine tracheas.

Some supplements purify out the chondroitin sulfates; others use the bovine trachea without further purification.

Functions

Chondroitin sulfate is a normal constituent of joint cartilage. Supplementation with this substance results in decreased activity of destructive joint enzymes. Chondroitin sulfate probably binds with these enzymes and inactivates them, protecting the joint cartilage. Chondroitin sulfate may also stimulate the production of hyaluronic acid, the substance which makes joint fluid thick and protective. You may encounter statements that chondroitin supplementation can protect joints/prevent problems. However, I have personally seen many cases of horses on manufacturer-recommended doses of chondroitin products who went on to develop new or worsened joint, ligament and tendon problems while taking the supplement. On the other hand, dosages recommended above are much higher than those often recommended by manufacturers so dosage may have played a role in the failure to protect these horses. Estimates of how much of chondroitin sulfate is actually absorbed intact range from 0 to 13%. One Japanese study reported about a 30% absorption rate. The remainder of the chondroitin is partially digested into small pieces which are then absorbed. Further breakdown occurs, probably in the liver for the most part, after these substances reach the blood stream. However, the benefits of chondroitin sulfate are not necessarily dependent on having it absorbed intact. Breakdown products of chondroitin may be just as, or possibly even more, active.

Supplementation

Feed at a rate of 7.5 grams/day to begin therapy. Dosage may be decreased after response is obtained. Drop dose slowly (by 1 to 2 grams per day) to 3.5 grams per day for maintenance. May increase again if symptoms worsen or horse is to be used for heavier work.

Deficiencies

Dietary deficiency per se cannot occur. However, arthritis and other inflammatory conditions of the joints may result in demand for precursors of joint cartilage that exceeds the body's ability to manufacture them.

Toxicity

None known. May cause digestive upset in sensitive horses.

Interactions

Complements the joint protective and rebuilding effects of chondroitin sulfates, glucosamine, manganese, vitamin C, copper and S-adenosylmethionine.

Indications

As a supportive therapy in the treatment of arthritis and ligament/tendon injuries.

CHROMIUM

PROBABILITY OF BENEFIT	DOSAGE	OVERDOSE RISK	COMPLEMENTARY NUTRIENTS	INDICATED FOR
	600-1,200 mcg	Unknown	Zinc Carbohydrates	• Building lean muscle mass
	600-1,200 mcg			• Controlling blood sugar
	1,200 mcg			• Encourage glycogen storage
	1,200-1,600 mcg			• Increase availability of fat as energy source
	1,200-1,600 mcg			

Sources

Chromium is a trace mineral present in all foods. Most effective form is the synthesized supplement chromium picolinate.

Functions

Chromium assists in the regulation of blood sugar, working with insulin. It is an integral component of the glucose tolerance factor in blood, which is involved with both processing glucose and keeping blood glucose levels stable. Supplementing chromium has been shown to decrease the amount of calories that are deposited as fat and increase lean muscle mass.

Supplementation

Precise recommendations for supplementing horses are unknown as yet. The above numbers are based on information available for other species and personal experience. Dosages as high as 5,000 mcg (5 mg) have been used in the experimental setting without adverse effects.

Deficiencies

None recognized but there is likely an optimal amount for horses (i.e., an amount that leads to maximal health and performance). Active horses may have a higher requirement for chromium, caused by increased excretion, or an increased sensitivity to its effects. Increasing intensity of exercise likely increases need.

Toxicity

Toxic dose has not been established but overuse of any mineral can lead to interference with absorption of others. Exceeding recommendations above is not recommended, not likely to be helpful and could be harmful.

Interactions

Chromium is essential to the normal functioning of insulin.

Indications

Chromium is currently found in some prerace carbohydrate loading formulas. It may be useful in the management of horses with pituitary tumors having problems with decreased insulin production. Chromium supplementation is strongly recommended for horses of any age receiving grain in their diets and whenever carbohydrate supplements are used. Improved muscle development has been seen in growing animals of other species. A study done on Thoroughbred racehorses in active training showed chromium decreased the insulin response to a meal (by increasing the cells' sensitivity to insulin, therefore decreasing the amount of insulin needed), decreased cortisol response during and after exercise (degree of cortisol release is directly related to insulin level at time of exercise), elevated blood triglycerides (the fat form used for energy) during and after exercise and led to a decreased production of lactate during moderately intense exercise. Decreased lactate production may have been related to improved utilization of triglycerides. All the above effects benefit exercising horses, particularly those exercising at the aerobic level (e.g., endurance horses). However, similar results were not obtained when testing horses that had not previously been in a training program. This is believed to reflect increased loss of chromium in exercising horses.

COENZYME Q$_{10}$

PROBABILITY OF BENEFIT	DOSAGE	OVERDOSE RISK	COMPLEMENTARY NUTRIENTS	INDICATED FOR
	180-300+ mg/day		Antioxidants Carnitine DMG Lipoic acid	• Lung disease • Heart disease • Old age • Exercise intolerance • Heavy exercise • Gum disease

Sources

The richest natural source of CoQ$_{10}$ in substances commonly given to horses is polyunsaturated vegetable oils. However, these oils must be cold processed or raw. Oils you buy in the grocery store have lost much of their nutritional value because they are processed to make them stable to use in cooking. An exception is extra virgin olive oil. The richest natural vegetable source of CoQ$_{10}$ is spinach—maybe CoQ$_{10}$ was Popeye's secret.

Functions

Exposure to drugs, chemicals, preservatives and inhaled impurities in the air result in the generation of substances called free radicals—electrically imbalanced molecules that attack normal body tissues to steal an electrical charge that will restore them to a normal state. The damaged cell in turn attacks its neighbors in the same way, setting up a chain reaction of cellular damage. The healthy body is constantly fighting off invasion by a large variety of bacteria, viruses and other organisms. When the immune system carries out this function, a normal waste product is the generation of free radicals which are potentially as damaging as any outside threat. Free radicals are believed to be largely responsible for many of the very familiar and uncomfortable symptoms of infection/inflammation (pain, swelling) and viral infections. Exercise and the generation of energy also results in free radical production that damages the exercising tissue. This process may cause the muscle aching and fatigue that follows heavy exercise. Antioxidants are substances present in the diet and manufactured by the body whose function is to neutralize free radicals before they can damage normal body tissues. If the production of free radicals exceeds the supply of antioxidants, tissue damage will result. CoQ$_{10}$ is an extremely potent antioxidant with a broad range of activities. CoQ$_{10}$ also improves immunity to bacteria and viruses and has been shown in research in people and other animals to reverse the declining immunity seen with aging. CoQ$_{10}$ is an extremely effective treatment for gum disease—a problem which plagues all older animals—and eliminates the need for surgery or costly antibiotics. Like the antioxidant lipoic acid, CoQ$_{10}$ also has an important role in energy generation. A syndrome has been described in people who have a genetic defect that leaves them with only 25% of the normal level of CoQ$_{10}$ in their bodies. The symptoms include seizures (the brain requires normal generation of energy to function properly) and—of particular interest to horse owners—muscular problems including decreased exercise tolerance, increased lactic acid in the blood, elevated muscle enzymes (CPK) and discolored urine after exercise (caused by the release of muscle pigment, myoglobin). We also know from studies in human athletes that CoQ$_{10}$ supplementation can improve athletic performance, at least in distance runners.

Supplementation

Antioxidants are a perfect example of how minimal intake of a nutrient may not be fatal but matching intake to needs results in greatly improved health. Situations which call for increased intake of antioxidants include infections, wounds, injuries such as sprains and strains, stressful situations such as shipping or change in environment, living in a polluted environment and, especially, exercise. No guidelines exist for CoQ_{10} supplementation in horses. The dosages suggested above are taken from work on humans and other animals.

Deficiencies

General signs of suboptimal antioxidant intake include poor stress tolerance, exercise related problems including subpar performance, frequent infections, poor wound healing. Over the long term, suboptimal antioxidant intake is believed to contribute to premature aging, cancer and health problems such as heart disease and diseases of the blood vessels. It is well documented in people that CoQ_{10} synthesis decreases with aging. With specific reference to CoQ_{10}, problems related to insufficient supply could include impaired immunity, gum disease, decreased exercise tolerance and muscle problems such as pain, cramping and even tying-up.

Toxicity

With the exception of vitamin A, antioxidants even in very large doses are basically nontoxic. There is no known toxicity associated with CoQ_{10} use.

Interactions

Many antioxidants complement the actions of other antioxidants. CoQ_{10} would complement the functions of all antioxidants with particular reference to lipoic acid, vitamin E and selenium.

Indications

Any horse can probably benefit from supplementation with antioxidants in terms of fewer infections, better wound healing, better stress tolerance, improved ability to detoxify drugs and other chemicals and possibly even a longer life. CoQ_{10} can do all of these things and is also extremely important to the efficient generation of energy in performance horses.

CREATINE

PROBABILITY OF BENEFIT PER USE	DOSAGE	OVERDOSE RISK	COMPLEMENTARY NUTRIENTS	INDICATED FOR
	12 grams daily after work or 60 grams daily for 3 days		Carbohydrate loading DMG	• Speed

Sources

Raw meat. Sold as a purified creatine powder or stabilized liquid.

Functions

Creatine is an amino acid, which is part of the compound creatine phosphate. It is stored in muscles as an immediately available source of energy, called upon particularly when speed is needed. Storage is limited and rapidly depleted.

Supplementation

Use only pure creatine powder or stabilized liquid. Must be administered separately and within less than 15 minutes of being exposed to any moisture, or the creatine degrades and is worthless. Do not mix large amounts in feed. You may suspend in small amount of thinned honey (honey and water), corn syrup or sugar water and give using a syringe. The sweets/sugars cause an insulin release which helps drive the creatine into the muscle cells. Note: No advantage to using larger amounts. Muscle has a limited capacity to store the creatine. Any extra will be excreted.

Deficiencies

Not applicable—true deficiency probably does not exist and would almost certainly be fatal. The horse's natural vegetarian diet may mean there is room in the cells for more creatine than is normally available from the diet or synthesis but this has not been proven as yet.

Toxicity

None known. Overdose of any amino acid may interfere with absorption of others. Creatine may result in intestinal discomfort, refusal of grain and bloating in sensitive horses.

Interactions

Acts independently in the muscle cell. Complementary supplements include those that target the systems responsible for producing speed.

Indications

To improve availability of short bursts of speed (sprints, gate speed, etc.). Allows horse to hold his maximum speed for a longer period of time but does not actually increase the horse's maximum speed. Allows horse to perform repetitive works requiring speed more easily (intervals, heats, etc.). Effect is most pronounced when exercise calls for repeated efforts over a short period of time (e.g., interval training). Effectiveness in horses is not definitely proven as yet. It would theoretically be of most benefit to horses performing at speed in multiple classes on the same day but interest is greatest in racing circles. As with most performance enhancing products, expect more effect in untrained horses than in those at peak fitness.

Storage and Handling

Store in dark, tightly sealed container at room temperature, low humidity. Do not expose to liquid until immediately prior to administering.

CYSTINE

AMINO ACID

PROBABILITY OF BENEFIT	DOSAGE	OVERDOSE RISK	COMPLEMENTARY NUTRIENTS	INDICATED FOR
	Not established		Methionine Balanced protein B-6	• Skin and coat problems • Poor wound healing • Feet • Joint and tendon problems

Sources

Protein is available from all common feeds. Fresh grasses and alfalfa hay contain the most, followed by grains and grass hays. The horse's body can manufacture some amino acids. Cystine is probably a nonessential amino acid in the horse that can be synthesized from methionine.

Functions

Amino acids are the building blocks of protein. There are essential and non-essential amino acids—essential meaning they must be present in the diet because the body cannot make them and nonessential meaning they can be manufactured by the horse. The essential amino acids for people are: threonine, lysine, valine, leucine, methionine, isoleucine, tryptophan, phenylalanine and histidine. In the horse, only lysine has been demonstrated to be essential—largely because it is the only one that has been studied. There is considerable interest currently in methionine and threonine but not enough information is yet available to advise on required dietary levels. The amino acids leucine, isoleucine and valine are of interest to those who work with high performance horses. Cystine and methionine can be interconverted in the body. Cystine is a powerful antioxidant, used in other species to treat respiratory problems, such as chronic bronchitis. It is important to the healing of wounds and burns and to the general health of the skin and hair (skin and hair are 10% to 14% cystine). Cystine also improves the immune response.

Supplementation

There are no guidelines for cystine supplementation in the diet. It is usually included in the methionine requirement when amino acid requirements are given for other species. We do not know the specific methionine requirement of horses either but we are fairly certain there is one. In other animals, methionine is required at a rate of 2 to 4% of the crude protein in the diet. Nutritionists often borrow information from pig nutrition to apply to the horse since pigs have a similar digestive tract. Pigs require 1.9% combined total methionine and cystine (another sulfur containing amino acid) in their diet. Supplementation at a rate of 1,500 to 2,000 mg per day is usually recommended.

Deficiencies

Deficiency of any of the essential amino acids will lead to poor growth and poor tolerance to stress. Specific symptoms for an isolated amino acid will depend upon those organ systems where it is particularly important. We have no specific deficiency information for horses concerning cystine but would expect to see poor quality skin and hair and poor wound healing.

Toxicity

Oversupplementation with cystine can decrease copper absorption and it is best to meet the needs with adequate methionine levels.

Interactions

There are many complex interactions between amino acids and other amino acids. The proper functioning of amino acids is also closely tied to adequate energy content in the diet and normal vitamin and mineral levels. Adequate B6 is needed to properly use proteins.

Indications

No specific recommendations. Short term supplementation may be beneficial in the case of severe wounds or burns.

DMG (DIMETHYLGLYCINE)/VITAMIN B-15
TMG (TRIMETHYLGLYCINE)/BETAINE

PROBABILITY OF BENEFIT	DOSAGE	OVERDOSE RISK	COMPLEMENTARY NUTRIENTS	INDICATED FOR
	1,500 to 3,000 mg		Carnitine Lipoic acid B vitamins CoQ_{10}	Improved aerobic exercise capacity in unfit horses

Sources

DMG is a substance that occurs naturally in all animals and plants. It has also been called vitamin B15. TMG is identical except that is has one more methyl group in its structure. TMG readily converts to DMG after it is eaten.

Functions

DMG and TMG are methyl donors which simply means that they have a methyl chemical group in their structure that is used to participate in a wide variety of reactions in the body. DMG has been shown to be an antioxidant and is involved in energy production processes that use oxygen (aerobic work) as well as the immune response.

Supplementation

The use of DMG in human athletes is based on the finding that the Russian teams which overwhelmingly dominated the 1960 Olympics were using a supplement that contained a large amount of DMG. However, research or even anecdotal evidence to support it as a miracle supplement have been lacking. There is one study showing that trimethylglycine can result in some improvement in aerobic work in horses that are untrained. Horses that were fit for the work level showed no benefit. The use of DMG to prevent tying-up is widespread but there is no solid evidence that it helps. However, horses prone to this problem that are likely to be asked to perform beyond their level of fitness may benefit. DMG or TMG are also sometimes used in the treatment of arthritis, for their antioxidant properties (they help generate the potent tissue antioxidant glutathione), generation of the anti-inflammatory SAM—S-adenosylmethionine—from methionine and generation of keratan, a glycosaminoglycan found in joint tissue as well as skin, hoof, hair.

Deficiencies

None known.

Toxicity

None known.

Interactions

Any nutrient that improves the efficiency of aerobic generation of energy and/or the use of fuels needed for aerobic generation of energy is complementary.

Indications

To improve aerobic generation of energy in horses that are not yet fit for the work level being asked of them.

ESSENTIAL FATTY ACIDS
LINOLEIC ACID, LINOLENIC ACID

PROBABILITY OF BENEFIT	DOSAGE	OVERDOSE RISK	COMPLEMENTARY NUTRIENTS	INDICATED FOR
	3 to 6 tablespoons per day		Antioxidants	• Control of inflammatory conditions • Brilliant coat • Production of hormones

Sources

Essential fatty acids are those that the body cannot manufacture and must be obtained from the diet. Essential fatty acids are present in unprocessed vegetable oils but oils that have been dehydrogenated, bleached, deodorized or heated have had the chemical shape of the oils altered to an extent that the body can no longer use them. *All grocery store shelf oils, except for extra virgin olive oils, have been processed.* The following is a list of vegetable oils and the essential fatty acid percentages they contain.

	LINOLEIC	LINOLENIC
Hempseed	58	20
Flaxseed	16.7	55
Pumpkin seed	45	15
Soybean	40	11
Walnut	50	5
Canola	26	8
Almond	17	
Virgin Olive	12	
Safflower	70	
Sunflower	66	
Corn	59	
Sesame	45	
Rice Bran	35	

Functions

The horse requires a small amount of essential fatty acids to carry out the business of producing adrenal gland and sex hormones, maintaining the structure of cell walls and in the function of the brain, eyes and ears. Proper levels of essential fatty acids are also important in regulating inflammatory responses in such conditions as chronic arthritis.

Supplementation

Horses with access to pasture and/or grains will probably get enough of the essential fatty acids in their diet, at least for maintenance (no work) purposes. Horses with skin and hoof problems will probably benefit from the addition of a fat source containing these fatty acids, as may horses with arthritis. Because the body may lose its ability to convert linoleic to linolenic and vice versa with age, older horses should receive a fat source that contains both.

Deficiencies

Likely signs of deficiency include dry, dull hair coat, hoof problems and difficult to control inflammatory conditions such as arthritis.

Toxicity

None known.

Interactions

For specific problems (e.g., skin), see Chapter 6, Relieving Health Problems Through Nutrition for other nutrients that may also be involved.

Indications

Chronic inflammatory conditions (arthritis, tendonitis), hoof and skin problems. General nutritional support of older horses.

FAT (VEGETABLE OR ANIMAL ORIGIN)

PROBABILITY OF BENEFIT	DOSAGE	OVERDOSE RISK	COMPLEMENTARY NUTRIENTS	INDICATED FOR
	5 to 15% of diet		Carnitine B vitamins Inositol Lipoic acid DMG CoQ_{10}	• Increased calories to maintain weight • Energy source for aerobic exercise • Increasing shine of coat

Sources

Fat is present in the basic diet of horses in very small amounts. The amount actually required for health is quite small.

Functions

Fats are only really needed to maintain normal hormone production and keep the production of good and bad messenger systems (such as the prostaglandins, which control such things as inflammation) in balance. Degenerative/inflammatory conditions such as arthritis may worsen in diets that do not contain sufficient amounts of essential fatty acids. Fats are an excellent source of calories. Pound per pound, fats provide 2.5 times the calories of carbohydrates or protein. Fats are therefore often used for horses that have difficulty gaining or holding weight. The digestion of fats produces less heat than digestion of carbohydrates, a fact that some consider of benefit in feeding during the summer months. Fats are a readily available muscle fuel for aerobic (slow) exercise. Essential fatty acids are specific types of fat that the horse's body requires but cannot manufacture on its own. Essential fatty acids are high in vegetable oils that have not been processed to withstand heating and have a long shelf life, e.g., raw linseed oil. Other vegetable fats are also good sources of essential fatty acids, but not if they have been processed as above. Fats you buy in the grocery store have had their essential fatty acids essentially destroyed by processing. It is possible to purchase vegetable oils that have not been degummed and processed from some animal supply houses, bulk suppliers of fat and other feedstuffs and bulk suppliers of feeds and vitamin/mineral additives for feeds. Fats that are high in the essential fatty acids are often fed at a rate of several tablespoons per day to improve the shine of the coat.

Supplementation

For weight gain purposes, liquid vegetable fats or processed animal fats (any fat supplement that is a solid/crystal/powder is animal fat—lard) are often used at a rate of anywhere from several ounces to 2 cups or more daily. Fats are used as a total replacement for grain in the treatment of polysaccharide storage disease—a specific defect of muscle which can cause tying-up. Fats are a common component of the diet for endurance horses, who have difficulty taking in enough hay and grain to hold their body condition during heavy training and competition. Fats may be used in the diet of other high performance horses as well, for the same reasons. However, feeding fat at high rates (over 10% of the total calories consumed) can result in the enzyme generating systems inside the muscle cell changing toward a profile that favors the use of fats for fuel. A decrease in the amount of stored glycogen in the muscle has also been reported with high fat feeding. This is fine as long as the horse is doing most or all of his work at aerobic levels. If high speed is required, high fat feeding may be a drawback.

Deficiencies

Deficiencies of essential fatty acids may occur on low fat and all hay diets, particularly as the animal ages. With age, the enzyme systems that are able to convert one form of essential fatty acid to another become less effective.

Toxicity

No direct toxicity of high fat feeding has been described as yet although the long term consequences of feeding lard (animal fat) to horses are not known. As above, high fat feeding may have a negative effect on speed. Feeding fat at a level of greater than 20 percent will likely cause the horse to go off feed (not palatable) and will cause softening of manure. Sensitive horses may develop abdominal distention and increased gas production if fat is introduced too quickly.

Interactions

Energy generation and proper use of fats is enhanced by those nutrients directly involved in fat metabolism as well as those that favor aerobic energy production.

Indications

Flax, hemp or raw vegetable oils would probably benefit the older horse and should be used on a daily basis. Fats also improve coat shine and quality. Provision of fat sources high in essential fatty acids may assist with arthritis control. Feeding vegetable or animal fat supplies a readily digestible and concentrated source of calories for horses with trouble gaining weight and/or horses performing a high level of aerobic/endurance exercise.

GAMMA ORYZANOL

PROBABILITY OF BENEFIT	DOSAGE	OVERDOSE RISK	COMPLEMENTARY NUTRIENTS	INDICATED FOR
	1,000 mg		Chromium BCAAs B6 Arginine	• Improved muscle mass • Improved stress tolerance • Improved fertility

Sources

Highest natural source is rice bran and also present in yams but effective dosing requires use of the purified gamma oryzanol.

Functions

Gamma oryzanol is a plant sterol/hormone with anabolic properties in horses and other animals. Limited studies in laboratory animals have documented nitrogen retention is improved—which translated means the animals are putting on more muscle. Clinical studies in horses by this author and other veterinarians have clearly shown stressed horses and horses in heavy training show improved muscling and progress through training much easier when given this product. Information is also emerging showing gamma oryzanol is helpful in getting problem mares into foal and maintaining good sterility in stallions.

Supplementation

Several gamma oryzanol products are on the market at this time but they are not all equally effective and some do not even contain the advertised amounts of gamma oryzanol. Liquid gamma oryzanol in fat emulsion suspension is the recommended form of this supplement. A controlled trial performed by this author comparing the fat emulsion form and various powdered gamma oryzanols showed no benefit from the powders but dramatic results with the fat emulsion form. Gamma oryzanol is a safe and very effective alternative to injectable anabolic steroids such as Equipoise or Winstrol—getting equivalent strengthening results with none of the harmful side effects.

Deficiencies

Not applicable.

Toxicity

None known. Appears safe for growing, pregnant, breeding and working animals.

Note

NOTE: Caution is probably indicated in using this product in growing animals until such time as beneficial growth and maturation effects have been proven on all body tissues. Rapid growth and weight gain could contribute to developmental bone and joint disease such as epiphysitis and OCD (osteochondrosis dessicans) unless bone/cartilage growth and strength is keeping pace. This author has no personal experience with using this product in growing horses although it has been done by many others. Anyone considering using gamma oryzanol in normal growing horses is encouraged to contact the manufacturer of the product for the latest in information on this topic.

Interactions

As an anabolic, gamma oryzanol allows you to get the most from your horse's diet. However, muscle cannot be built maximally if there is inadequate total protein intake or if protein is of poor quality. Complements other natural anabolics such as chromium and branched chain amino acids (BCAAs) used pre- and post-exercise. Vitamin B6 is essential for normal protein metabolism.

Indications

Maintaining and building muscle mass in performance animals. Improved fertility in breeding animals. Reducing negative effects of stress.

GLUCOSAMINE HYDROCHLORIDE

Glucosamine Sulfate

PROBABILITY OF BENEFIT	DOSAGE	OVERDOSE RISK	COMPLEMENTARY NUTRIENTS	INDICATED FOR
	9 gm/day		Chondroitin sulfate Perna mussel Copper Manganese MSM GAGs S-adenosylmethionine	• Arthritis • Tendinitis • Ligament injury • Old age • GI ulcers

Sources

Not present in normal equine diet.

Functions

Glucosamine forms the backbone of the GAGs (glycosaminoglycans), the building blocks of cartilage and all other connective tissues throughout the body. Glucosamine has been shown to be the rate-limiting step in the formation and healing of cartilage. In other words, without sufficient glucosamine production, these processes cannot keep up with damage. Providing glucosamine to joint cells stimulates them to produce more GAGs such as chondroitin sulfate. In high amounts, glucosamine also stimulates the products of hyaluronic acid, an important constituent of joint fluid. Because chondroitin levels in cartilage are increased when glucosamine is given, glucosamine can also have the enzyme inhibiting effects seen when chondroitin supplements are used.

Supplementation

Glucosamine is a small molecule and is absorbed intact at a rate of anywhere from 90 to 98%. Absorption of glucosamine is much easier for the intestinal cell than absorption of an intact chondroitin molecule. A comparison would be how easy it is for you to swallow whole a strand of spaghetti compared to a hair brush. There is considerable debate about whether glucosamine hydrochloride or glucosamine sulfate is the preferred form for a joint supplement. Both are very easily absorbed. The sulfated form contains less actual glucosamine per weight than does the hydrochloride. It is also believed that enzymes chop off the sulfate portion before absorbing preparations of glucosamine sulfate. On the other hand, glucosamine has been shown to work better in the presence of sulfur. For maximal effect you could combine glucosamine with a bioavailable source of sulfur such as methionine, S-adenosylmethionine or MSM. The 9 gram dosage is recommended initially, until benefit is obtained. After this, you can gradually (1 to 2 grams per day) decrease to a maintenance dose of 3 to 4.5 grams/day—more for very large horses. Increase as needed if symptoms recur or if heavier work than normal is anticipated.

Deficiencies

Dietary deficiency per se cannot occur. However, arthritis and other inflammatory conditions of the joints may result in demand for precursors of joint cartilage that exceeds the body's ability to manufacture them. It is also known that aging results in decreased production of glucosamine, as well as decreased production of all the GAGs formed from glucosamine.

Toxicity

None known.

Interactions

Complements the joint protective and rebuilding effects of chondroitin sulfates, glucosamine, manganese, vitamin C, copper and S-adenosylmethionine.

Indications

As a supportive therapy in the treatment of arthritis and ligament/tendon injuries.

GRAPE SEED EXTRACT

PROBABILITY OF BENEFIT	DOSAGE	OVERDOSE RISK	COMPLEMENTARY NUTRIENTS	INDICATED FOR
	0.6 to 1 mg/lb or as directed		Bioflavinoids Vitamin C Vitamin E Selenium Zinc Copper Lipoic Acid CoQ_{10}	General • Antioxidant • Allergic reactions • Anti-aging • Stabilization of blood vessels • Protecting exercising muscles • Tumorous growths • Assist in control of lung bleeding

Sources

Extracted from seeds, skin and pulp of grapes. Highest concentration in red grapes.

Functions

Grape/grape seed extract contains many active antioxidant chemicals, including bioflavinoids and polyphenols such as proanthocyanidin. The latter in particular have extremely potent antioxidant activity, being from 20 to 50 times more active than vitamin C and vitamin E. Grape seed extract also assists the entry of vitamin C into cells. It helps to stabilize histamine release, making it a natural approach to allergy control. Grape seed extract protects small blood vessels from damage to their walls, including the capillaries in the lung. The anti-inflammatory/antioxidant effects have led to its use to assist in arthritis control, relief from allergy symptoms (including breathing problems caused at least in part by allergies) and help prevent damage to heavily exercising muscles. Anticancer/antitumor and general anti-aging benefits have also been proposed.

Supplementation

Usual dosage is 0.6 to 1.0 mg/lb. Higher doses may be used for horses with specific problems that may benefit (e.g., lung bleeding, COPD/heaves, heavy exercise, etc.).

Deficiencies

Not applicable.

Toxicity

None known.

Interactions

Supports and complements any of the antioxidant vitamins, minerals or nutrients as listed above.

Indications

Helpful in controlling the common following equine conditions: Lung bleeding, allergic lung disease, arthritis, tendonitis, myositis, periodontal disease, chronic inflammatory conditions such as recurrent uveitis (moonblindness). Grape seed extract topical poultices have been used to shrink melanomas—pigment containing tumors common to older grey horses.

INOSINE (HYPOXANTHINE)

PROBABILITY OF BENEFIT	DOSAGE	OVERDOSE RISK	COMPLEMENTARY NUTRIENTS	INDICATED FOR
	5 grams		None	No clear indications

Sources

Inosine is present in all living tissues but is synthesized for supplements.

Functions

In the body, inosine is a waste product. It is formed when the cells break down their energy store, ATP. Inosine is normally processed to urea and excreted in the urine. The body does not reuse inosine in any significant amounts under normal circumstances.

Supplementation

Inosine shows up in some supplements sold to enhance performance but there is no evidence to support its use and some studies show harmful results (see Toxicity). Furthermore, absorption studies have clearly shown that inosine taken orally is immediately cap-tured by the cells of the intestinal tract and taken to the liver where it is recognized as a toxin/waste product, broken down and excreted. Inosine never makes it to the muscles in any significant amount.

Deficiencies

None.

Toxicity

Research in human athletes has shown inosine has the potential to negatively effect performance. No benefits from this substance have been shown in controlled studies or clinical trials.

Interactions

No beneficial interactions. Feeding inosine only adds to the burden of waste products that the body must remove.

Indications

None.

The interest in inosine as a performance aid arose when reports appeared that Russian and other Iron Curtain athletes were using this substance. It hit the health food stores before any studies into its benefit had been performed. There is research showing that when hearts are taken out of the body, continue to work and are deprived of all nutrition for several hours, they have the capacity to recycle inosine into a useful compound. However, that obviously does not apply to normal exercising horses, or any other whole animal or human. This author has seen horses lose speed when given inosine. This was not a controlled scientific study and included only a small number of horses—enough, however, to convince me never to use it again. If you want to try a product that contains inosine, I strongly suggest you try it out first during a heavy training session or a relatively unimportant race to see if you like the results.

ISOLEUCINE

BRANCHED CHAIN AMINO ACID

PROBABILITY OF BENEFIT	DOSAGE	OVERDOSE RISK	COMPLEMENTARY NUTRIENTS	INDICATED FOR
	30 grams*		B6 Balanced protein	• Tying-up • Muscle cramp/ pain • Fatigue • Decreased muscle mass • Elevated blood levels of muscle enzymes
	*Refers to 30 grams of mixed branch chain amino acids, not isoleucine alone.			

Sources

Protein is available from all common feeds. Fresh grasses and alfalfa hay contain the most, followed by grains and grass hays. The horse's body can manufacture some amino acids. Branched chain amino acids occur in all foods but are highest in milk and meat proteins.

Functions

Amino acids are the building blocks of protein. There are essential and nonessential amino acids—essential meaning they must be present in the diet because the body cannot make them and nonessential meaning they can be manufactured. The essential amino acids for people are: threonine, lysine, valine, leucine, methionine, isoleucine, tryptophan, phenylalanine and histidine. In the horse, only lysine has been demonstrated to be essential. There is considerable interest currently in methionine and threonine but not enough information is available as yet to advise on required dietary levels. The amino acids leucine, isoleucine and valine are of interest to those who work with high performance horses. The latter are the branched chain amino acids. Branched chain amino acids are broken down in the muscles during exercise. This provides a source of energy. They also can use pyruvate generated during the breakdown of glycogen to regenerate alanine that is lost during exercise, resulting in a lower production of lactate. There is some evidence that BCAAs may also substitute for glycogen as an energy source during exercise. It is further theorized that by decreasing the concentration of tryptophan in the blood, BCAAs may help prevent the sensation of fatigue, although there are conflicting reports on this. BCAAs have been used to help prevent tying-up. Their use was associated with absence of symptoms and drop in blood levels of muscle enzymes with exercise. Most recipes use 30 grams of BCAAs and roughly equal amounts of valine and leucine, slightly less than half as much isoleucine. Isoleucine is needed for hemoglobin synthesis and is reported to stabilize and maintain blood sugar levels.

Supplementation

Supplementation may be helpful during aerobic exercise—prolonging the time to depletion of glycogen stores and sparing the BCAAs in the muscle from degradation (the muscle will take what it needs from blood if supplies are high enough). Some types of tying-up may benefit from treatment with BCAAs.

BCAAs should be administered approximately one half hour before and one half hour after exercise. For exercise bouts longer than two hours, repeat treatment during exercise may be indicated. Routine supplementation is not recommended.

Deficiencies

Deficiency of any of the essential amino acids will lead to poor growth and poor tolerance to stress. Specific symptoms for an isolated amino acid will depend upon those organ systems where it is particularly important. In the case of BCAAs, loss of muscle bulk and poor exercise tolerance may be seen. Deficiency of isoleucine has been reported in other species to cause symptoms similar to hypoglycemia (low blood sugar).

Toxicity

Excessive amounts may cause an abnormally low blood sugar.

Interactions

There are many complex interactions between amino acids and other amino acids. The proper functioning of amino acids is also closely tied to adequate energy content in the diet and normal vitamin and mineral levels. BCAAs restore levels of alanine and glutamine lost during exercise and have a sparing effect on muscle glycogen.

Indications

As above, BCAAs may be of benefit during aerobic exercise, particularly endurance events. Some horses with tying-up may also benefit by treatment with BCAAs. The effect of BCAA supplementation on speed performance requires more evaluation.

L-ARGININE

AMINO ACID

PROBABILITY OF BENEFIT	DOSAGE	OVERDOSE RISK	COMPLEMENTARY NUTRIENTS	INDICATED FOR
	12 gm per 100 kg* *See below for details.		Gamma oryzanol B-6	• Increased muscular mass and strength

Sources

Must be used as a purified single amino acid from a commercial source when supplementing.

Functions

L-arginine is used by human athletes to trigger the release of growth hormone from the brain. Growth hormone is a powerful stimulator of muscle, resulting in increased mass and strength with decreased body fat—in short, it is an anabolic. Release of growth hormone is triggered by exercise and is largely responsible for the improved muscle mass and strength resulting from training. Arginine has been proven to enhance growth hormone release in human athletes. Studies on horses are lacking.

Supplementation

Supplementation with arginine to get the growth hormone releasing effect is tricky. It must be given as the pure amino acid and recommendations suggest giving on a completely empty stomach and several hours distant from the intake of any other protein source. In fact, some human athletes get up in the middle of the night to take their arginine! L-arginine should NOT be given on days the horse does not exercise and should be used in cycles of no more than 12 weeks on the supplement with 6 to 8 weeks off. L-arginine tastes terrible. You will have to dissolve it in a small amount of water and administer by dose syringe. There are L-arginine products for horses on the market. Recommended dosage is usually 30 grams/day. However, effective dose in people is 12 gm/100 kg which would translate into 60 gm/day for a 500 kg (1,100 pound) horse.

Deficiencies

There is no evidence that a deficiency of the amino acid arginine exists on normal diets.

Toxicity

Side effects of headache, nausea and diarrhea have been reported in humans and are certainly theoretically possible with horses. Any horse showing signs of abdominal pain, depression, irritability or any other type of adverse reaction after dosing should not be given L-arginine. Use in horses is largely uncharted territory and you will be proceeding at your own risk if you try this supplement. L-arginine can also cause activation of latent virus in the body in humans (e.g., herpes outbreak—cold sores or genital). Long term safety has not been established in any species.

Interactions

Use of gamma oryzanol at least theoretically would complement L-arginine. However, the effectiveness and safety of this combination is unknown. The amino acids L-ornithine alpha-ketoglutarate and glycine have also been used in humans for a growth hormone releasing effect. Equivalent equine doses would be in the range of 40 gm/day of glycine and 6 gm/100 kg/day of L-ornithine alpha-ketoglutarate. Again, the effectiveness and safety of trying this in horses is unknown at this time.

Indications

To assist in building muscle mass and strength during training.

LEUCINE

BRANCHED CHAIN AMINO ACID

PROBABILITY OF BENEFIT	DOSAGE	OVERDOSE RISK	COMPLEMENTARY NUTRIENTS	INDICATED FOR
	30 grams*		B-6 Balanced protein	• Tying-up • Muscle cramp/pain • Fatigue • Decreased muscle mass • Elevated blood levels of muscle enzymes

*Refers to 30 grams of mixed branched chain amino acids, not leucine alone.

Sources

Protein is available from all common feeds. Fresh grasses and alfalfa hay contain the most, followed by grains and grass hays. The horse's body can manufacture some amino acids. Branched chain amino acids occur in all foods but are highest in milk and meat proteins.

Functions

Amino acids are the building blocks of protein. There are essential and nonessential amino acids—essential meaning they must be present in the diet because the body cannot make them and nonessential meaning they can be manufactured. The essential amino acids for people are: threonine, lysine, valine, leucine, methionine, isoleucine, tryptophan, phenylalanine and histidine. In the horse, only lysine has been demonstrated to be essential. There is considerable interest currently in methionine and threonine but not enough information is available as yet to advise on required dietary levels. The amino acids leucine, isoleucine and valine are of interest to those who work with high performance horses. The latter are the branched chain amino acids. Branched chain amino acids are broken down in the muscles during exercise. This provides a source of energy. They also can use pyruvate generated during the breakdown of glycogen to regenerate alanine that is lost during exercise, resulting in a lower production of lactate. There is some evidence that BCAAs may also substitute for glycogen as an energy source during exercise. It is further theorized that by decreasing the concentration of tryptophan in the blood, BCAAs may help prevent the sensation of fatigue, although there are conflicting reports on this. BCAAs have been used to help prevent tying-up. Their use was associated with absence of symptoms and drop in blood levels of muscle enzymes with exercise. Most recipes use 30 grams of BCAAs and roughly equal amounts of valine and leucine, slightly less than half as much isoleucine. Leucine also is reported to lower blood sugar and aid in healing of wounds and fractures.

Supplementation

Supplementation may be helpful during aerobic exercise—prolonging the time to depletion of glycogen stores and sparing the BCAAs in the muscle from degradation (the muscle will take what it needs from blood if supplies are high enough). Some types of tying-up may benefit from treatment with BCAAs.

BCAAs should be administered approximately one half hour before and one half hour after exercise. For exercise bouts longer than two hours, repeat treatment during exercise may be indicated. Routine supplementation is not recommended.

Deficiencies

Deficiency of any of the essential amino acids will lead to poor growth and poor tolerance to stress. Specific symptoms for an isolated amino acid will depend upon those organ systems where it is particularly important. In the case of BCAAs, loss of muscle bulk and poor exercise tolerance may be seen.

Toxicity

Excessive amounts may cause an abnormally low blood sugar.

Interactions

There are many complex interactions between amino acids and other amino acids. The proper functioning of amino acids is also closely tied to adequate energy content in the diet and normal vitamin and mineral levels. BCAAs restore levels of alanine and glutamine lost during exercise and have a sparing effect on muscle glycogen.

Indications

As above, BCAAs may be of benefit during aerobic exercise, particularly endurance events. Some horses with tying-up may also benefit by treatment with BCAAs.

LIPOIC ACID

ANTIOXIDANT

PROBABILITY OF BENEFIT	DOSAGE	OVERDOSE RISK	COMPLEMENTARY NUTRIENTS	INDICATED FOR
	300-1,200 mg/day		Vitamin C Vitamin E Carnitine	• Antioxidant protection • Tying-up • Heavy exercise

Sources

While lipoic acid is present to some extent in all foods, it is found in the highest concentration in red meats, not the vegetarian diet a horse consumes. The body manufactures lipoic acid.

Functions

Exposure to drugs, chemicals, preservatives and inhaled impurities in the air result in the generation of substances called free radicals, electrically imbalanced molecules that attack normal body tissues to steal an electrical charge that will restore them to a normal, neutral state. The damaged, and now charged, cell in turn attacks its neighbors in the same way, setting up a chain reaction of cellular damage. The healthy body is constantly fighting off invasion by a large variety of bacteria, viruses and other organisms. When the immune system carries out this function, a normal waste product is the generation of free radicals which are potentially as damaging as any outside threat. Free radicals may be responsible for many of the very familiar and uncomfortable symptoms of infection/inflammation (pain, swelling) and viral infections. Exercise and the generation of energy also result in free radical production that damages the exercising tissue. This process is considered to be a significant cause of the muscle aching and fatigue that follows heavy exercise. Antioxidants are substances present in the diet and manufactured by the body whose function is to neutralize free radicals before they can damage normal body tissues. If the production of free radicals exceeds the supply of antioxidants, tissue damage will result. Lipoic acid is an extremely potent antioxidant, more than vitamin E or vitamin C. Another distinct advantage of lipoic acid is that it can go into both cell walls and other fatty structures (where vitamin E works) and into the fluids of the cells and blood (vitamin C's territory). Lipoic acid can protect from free radical attack anywhere in the body, on any type of structure. Lipoic acid also is capable of restoring itself to an active form again after it has taken on a free radical, something the other antioxidants cannot do, and will perform the same function for vitamin C or vitamin E. Lipoic acid also has very important functions in the generation of energy. Lipoic acid assists in getting glucose inside the cells where it is needed (in humans and dogs it is used to treat diabetes and insulin resistance—to get the blood sugar inside the cells where it belongs). Having done that, lipoic acid also works to get that energy source burned efficiently.

Supplementation

Minimal intake of antioxidants may not be fatal but matching intake to needs results in greatly improved health. Situations which call for increased intake of antioxidants include infections, wounds, injuries such as sprains and strains, stressful situations such as shipping or change in environment, living in a polluted environment and, especially, exercise. You will not find any discussions of lipoic acid in the NRC Guidelines for horses and no recommendation for dietary level or supplementation exists. The dosages given above reflect recommended intake for humans and other animals, adapted for horses. Lipoic acid can be found in a few supplements manufactured for high performance horses and/or horses with muscle problems and tying-up. You can see from the description of its functions that it is an entirely reasonable ingredient there. However, if you are going to give lipoic acid a trial you must realize that as a nutrient it will require time to work and giving it on a one time only basis, just a few hours before a performance, will have minimal effects compared to using it for a few days before the race, hunt, show or whatever. It normally takes weeks of supplementation before you can accurately identify any benefits from a nutrient.

Deficiencies

General signs of suboptimal antioxidant intake include poor stress tolerance, exercise related problems including subpar performance, frequent infections, poor wound healing. Over the long term, suboptimal antioxidant intake is believed to contribute to premature aging, cancer and health problems such as heart disease and diseases of the blood vessels. In the case of lipoic acid you can add decreased performance (e.g., less speed, finishing poorly in a race) and muscle problems of pain, cramping even tying-up to the list.

Toxicity

With the exception of vitamin A, antioxidants even in very large doses are basically nontoxic. Because of lipoic acid's effect of helping to drive blood glucose into the cells, it lowers insulin requirements in diabetics and high doses can cause blood sugar to drop too low. Whether this can occur in horses is unknown. If you are using lipoic acid containing supplements for only one or two days at a time, it is not likely to be a problem. However, if you plan to use it daily for longer than that, it is advisable to stay with the lower dosage ranges.

Interactions

Many of the antioxidants complement the actions of one or more other types of antioxidants. Lipoic acid can perform the functions of any of the other antioxidants and its use may lower requirements for such things as vitamin C and vitamin E. Lipoic acid also recycles vitamin E and vitamin C back to active forms. Because of its effects on glucose transport and metabolism, it supports the function of insulin and chromium in getting glucose into the cells and enhances the efficiency of energy generating pathways.

Indications

Any horse can probably benefit from supplementation with antioxidants in terms of fewer infections, better wound healing, better stress tolerance, improved ability to detoxify drugs and other chemicals and possibly even a longer life. Lipoic acid supplementation would be particularly indicated for high performance horses and horses with high energy requirements.

METHIONINE

AMINO ACID

PROBABILITY OF BENEFIT	DOSAGE	OVERDOSE RISK	COMPLEMENTARY NUTRIENTS	INDICATED FOR
	1.5 to 2.0 gm/day		B6 Balanced Protein Cystine	• Hoof quality problems • Skin, coat, tendon and ligament problems • Arthritis • Poor adaptation to and tolerance of exercise

Sources

Protein is available from all common feeds. Fresh grasses and alfalfa hay contain the most, followed by grains and grass hays. The horse's body can manufacture some amino acids. With specific reference to methionine, data on content in hays is not available. Corn and soybean meals are known to contain low levels of methionine and use of these feeds in other species where methionine requirements are better described requires that methionine be supplemented.

Functions

Amino acids are the building blocks of protein. There are essential and nonessential amino acids—essential meaning they must be present in the diet because the body cannot make them and nonessential meaning they can be manufactured by the horse. The essential amino acids for people are: threonine, lysine, valine, leucine, methionine, isoleucine, tryptophan, phenylalanine and histidine. In the horse, only lysine has been demonstrated to be essential—largely because it is the only one that has been studied. There is considerable interest currently in methionine and threonine but not enough information as yet to advise on required dietary levels. The amino acids leucine, isoleucine and valine are of interest to those who work with high performance horses. Methionine is a sulfur-containing amino acid and therefore is one of the essential sources of sulfur for the body (see sulfur). Methionine is important in protein metabolism and the generation of energy. It is also very important to the maintenance of normal skin, hooves, tendons, ligaments and cartilage.

Supplementation

We do not know the specific methionine requirement of horses but are fairly certain there is one. In other animals, methionine is required at a rate of 3 to 4% of the crude protein in the diet. Nutritionists often borrow information from pig nutrition to apply to the horse since pigs have a similar digestive tract. Pigs require 1.9% combined total methionine and cystine (another sulfur containing amino acid) in their diet. Supplementation at a rate of 1,500 to 2,000 mg per day is usually recommended in the treatment of conditions that may have inadequate methionine as part of their cause (e.g., poor hoof quality). Methionine also prevents fat buildup in the liver and helps detoxify chemicals. The horse's body uses methionine to create two other sulfur amino acids—taurine and cystine.

Deficiencies

Deficiency of any of the essential amino acids will lead to poor growth and poor tolerance to stress. Specific symptoms for an isolated amino acid will depend upon those organ systems where it is particularly important. We have no specific deficiency information for horses concerning methionine. It is suspected to play a role in poor hoof quality and may also be involved in tendon and ligament disease as well as poor adaptation to exercise.

Toxicity

Excessive methionine can interfere with the delicate balance of amino acids and may actually cause some problems you are trying to fix, such as poor growth rate. Methionine is associated with a higher incidence of side effects in humans including gas production, uneasy feeling and increased calcium excretion in the urine.

Interactions

There are many complex interactions between amino acids and other amino acids. The proper functioning of amino acids is also closely tied to adequate energy content in the diet and normal vitamin and mineral levels. The supply of cystine and taurine is dependent on adequate methionine. Adequate B6 is needed to properly utilize proteins.

Indications

Methionine supplementation may be indicated in the treatment of skin and hoof problems as well as to provide nutritional support to horses in high stress situations such as heavy exercise.

MSM (METHYLSULFONYLMETHANE)

PROBABILITY OF BENEFIT	DOSAGE	OVERDOSE RISK	COMPLEMENTARY NUTRIENTS	INDICATED FOR
	10 grams		Glucosamine Chondroitin sulfate Copper Vitamin C Methionine B6	• Arthritis • Tendonitis

Sources

MSM is a byproduct of the metabolism of DMSO that is found to exist naturally, including in the horse's body. MSM products are created from DMSO by enzymatic conversions. In nature, it is present in many fresh plants but is very volatile and susceptible to drying. Fresh alfalfa has detectable levels of MSM but alfalfa hay does not. Cow's milk, and presumably other milks, have high levels of MSM.

Functions

MSM serves as an organic source of sulfur. Sulfur is a mineral that is present in the body in small amounts but serves critical functions in the formation of production of cross-bridges/links between collagen molecules. These sulfur cross-bridges add strength and stability to the structure of collagen rich tendons, ligaments and joint tissues as well as to the connective tissues throughout the body. It has been shown that glucosamine, an important precursor of many different types of connective tissue, functions better in the presence of an adequate source of sulfur. Sulfur is also essential to the function of many hormones, enzymes and the immune system.

Supplementation

Usual dose is 10 grams/day, added to the feed. This is only an estimate and was arrived upon by determining how much MSM a 500 kg (1,100 pound) horse would take in from natural sources if being maintained on fresh pasture. Much higher doses (two to four times as much) have been used without ill effect.

Deficiencies

See sulfur in A to Z. Inadequate amounts of sulfur would be expected to lead to poor hoof strength, tendon/ligament weakness, poor joint cartilage quality. Repair of injury to any of these tissues would also be delayed/suboptimal. Studies in other species have also suggested MSM may be of benefit in the treatment of allergies, difficult to clear infections, parasite infestations, stress in general and intestinal tract ulcers.

Toxicity

None known in these dosage ranges. A toxic dose for rats has been established to be 20 gm/kg of body weight which would be 10,000 grams for a 500 kg horse.

Interactions

MSM can only help if there are adequate amounts of all the other nutrients required to build and repair connective tissue/hoof structures, including adequate protein, vitamin B6 to help process that protein, vitamin C, copper.

Indications

Under conditions of heavy stress/injury to joints, ligaments and tendons. Possible applications to problems with infections, allergies and intestinal parasites.

PERNA MUSSEL

PROBABILITY OF BENEFIT	DOSAGE	OVERDOSE RISK	COMPLEMENTARY NUTRIENTS	INDICATED FOR
	7 to 9 gm/day		Glucosamine Chondroitin sulfate Vitamin C MSM Manganese Copper	• Arthritis • Tendon and ligament problems • Old age

Sources

Not present in normal equine diet. Perna mussel is obtained from the sea. The entire organism is then freeze dried and included in equine supplements for joint disease.

Functions

Source of essential nutrients for joint health. Provides chondroitin sulfate as well as other glycosaminoglycans—i.e., other similar chemicals found in cartilage. Whether this has effects closer to that of chondroitin sulfate or a combination therapy of glucosamine and chondroitin sulfate is not clear. (See chondroitin sulfate and glucosamine for discussion of their effects.) Also a source of essential trace minerals and beneficial fatty acids. May also assist in healing of tendons and ligaments but this has not been proven.

Supplementation

Feed at rate of 7 to 9 grams/day to begin therapy. Dosage may be decreased after response is obtained. Drop dose slowly (by 1 to 2 grams per day) to 3.5 to 4.5 grams per day for maintenance. May increase again if symptoms worsen or horse is to be used for heavier work.

Deficiencies

Dietary deficiency per se cannot occur. However, arthritis and other inflammatory conditions of the joints may result in demand for precursors of joint cartilage that exceeds the body's ability to manufacture them.

Toxicity

None known. May cause digestive upset in sensitive horses.

Interactions

Complements the joint protective and rebuilding effects of chondroitin sulfates, glucosamine, manganese, vitamin C, copper.

Indications

As a supportive therapy in the treatment of arthritis and ligament/tendon injuries.

PHENYLALANINE

PROBABILITY OF BENEFIT	DOSAGE	OVERDOSE RISK	COMPLEMENTARY NUTRIENTS	INDICATED FOR
	3 to 9 grams		B6 Vitamin C	• Chronic pain

Sources

Phenylalanine is an amino acid that occurs naturally in foods.

Functions

Involved in the production of neurotransmitters such as norepinephrine, epinephrine and dopamine.

Supplementation

Phenylalanine has been suggested as a natural pain relieving treatment because of its proposed ability to elevate mood and block enzymes which break down endorphins—the body's natural pain killing and feel good chemicals. However, proponents warn it could take several weeks to see any effect (any pain could improve on its own over this time period) and recommend that other traditional pain relieving drugs be continued for enhanced effect. Fact is, it is likely the other drugs are responsible for all or most of the effect. Phenylalanine is not terribly effective.

Deficiencies

None known.

Toxicity

Side effects at high doses in people include headaches, irritability, insomnia and elevated blood pressure. Side effects in horses are unknown.

Interactions

B6 and vitamin C are required for conversion of phenylalanine to neurotransmitter chemicals.

Indications

If you are trying to avoid drugs in a horse with chronic pain such as arthritis, you could give phenylalanine a trial. However, you should do this at a time when the horse is receiving no other drug or nutritional therapy for the problem so that you can evaluate exactly what the effects of the phenylalanine are and avoid wasting your money if it does not work.

PROBIOTICS

PROBABILITY OF BENEFIT PER USE	DOSAGE	OVERDOSE RISK	COMPLEMENTARY NUTRIENTS	INDICATED FOR
	100,000+ of each organism	for adults	Yeast Fiber	• Improving digestion • Chronic low grade colic • Weight gain problems • Hay belly

Sources

Horses populate their intestinal tract with millions of organisms, picked up from the environment in general and exposure to manure of other horses. Probiotic products contain one or more specifically selected and grown beneficial strains of bacteria and/or growth factors that promote the health of beneficial bacteria in the intestinal tract.

Functions

To aid in digestion of coarse plant materials and assist in breakdown of nondigestible components of feeds. Healthy populations help protect from infection by dangerous bacteria such as salmonella. The end result is better digestion of feeds leading to decreased manure production, more firm manure, less gas and fluid accumulation in the bowel ("hay belly"), improved weight gain and improved utilization of all nutrients.

Supplementation

Products that contain a broad range of live bacteria (not too many do) or factors which enhance bacterial growth rather than the bacteria themselves are generally superior to those which only contain a few strains of bacteria.

Deficiencies

The bacterial populations of the intestine may be disrupted by many things including stress, shipping, change in diet, oral medications or wormers, ingestion of harmful bacteria, spoiled feed, inadequate water intake. In addition, anything that interferes with the normal motility and function of the intestine (such as damage from prior parasite burdens) may make it difficult for the beneficial bacteria to create a favorable environment.

Toxicity

Low to no possible toxicity in adults. Foals have died of iron overload/toxicity when given probiotic products that are high in iron. Do not give to very young foals (nursing only, not yet picking at feed) except on the direct order of a veterinarian. Can use yogurt made with live cultures in nursing foals.

Interactions

Yeast also improves some aspects of digestion and absorption of nutrients, but overfeeding yeast can cause digestive upset. Adequate fiber (hay) in the diet is essential to intestinal health and health of the beneficial bacteria.

Indications

To improve digestion and use of feeds, treat some cases of chronic, low-grade colic and improve weight gains.

SUPEROXIDE DISMUTASE (SOD)

PROBABILITY OF BENEFIT	DOSAGE	OVERDOSE RISK	COMPLEMENTARY NUTRIENTS	INDICATED FOR
	No effective dose		Zinc Copper Manganese All antioxidants	No indications for oral use

Sources

SOD is found in any living cell that utilizes oxygen. SOD supplements are synthetic forms of this enzyme.

Functions

SOD is an extremely important antioxidant enzyme. It functions (with its co-factors zinc, copper and manganese) to neutralize damaging free radicals of oxygen (waste products of body reactions using oxygen to generate energy or neutralize harmful chemicals and drugs). It is found almost exclusively inside the cells of the body (i.e., not freely floating in the blood).

Supplementation

There is absolutely NO evidence to suggest that giving SOD orally is of any benefit whatsoever. SOD is rapidly broken down by the digestive tract. Even injection of SOD directly into the blood stream is of very questionable benefit since the enzyme is destroyed and does not make its way inside the cells.

(The injectable anti-inflammatory Palosein, used by some veterinarians to treat arthritic joints, has SOD as its active ingredient.) SOD continues to appear on the ingredients list of several equine supplements, most notable those targeting arthritis.

Deficiencies

Less than optimal functioning of the SOD enzymes can occur whenever there is a deficiency of zinc, copper or manganese.

Toxicity

None known or likely.

Interactions

Complements the functions of all antioxidant systems in the body.

Indications

No indications for oral SOD.

Although giving this nutrient orally is a waste of money, SOD itself is a very important enzyme in the body. The problem is getting it into the cells where it is needed. Intra-articular (in the joint) injection of Palosein benefits some horses—most benefit being likely in cases that involve a great deal of recent, active inflammation rather than longstanding arthritic change. Very promising research is currently being performed on a liposomal form of SOD termed LIPSOD. This chemical has the SOD inside a protective envelope that is attracted to cell walls. Look for news on use of injectable LIPSOD in the future and perhaps future development of a similarly protected form that works orally.

VALINE

BRANCHED CHAIN AMINO ACID

PROBABILITY OF BENEFIT	DOSAGE	OVERDOSE RISK	COMPLEMENTARY NUTRIENTS	INDICATED FOR
	30 grams*		B6	• Tying-up • Muscle cramp/ pain • Fatigue • Decreased muscle mass • Elevated blood levels of muscle enzymes
	*Refers to 30 grams of mixed branched chain amino acids, not valine alone.			

Sources

Protein is available from all common feeds. Fresh grasses and alfalfa hay contain the most, followed by grains and grass hays. The horse's body can manufacture some amino acids. Branched chain amino acids occur in all foods but are highest in milk and meat proteins.

Functions

Amino acids are the building blocks of protein. There are essential and nonessential amino acids—essential meaning they must be present in the diet because the body cannot make them and nonessential meaning they can be manufactured by the horse. The essential amino acids for people are: threonine, lysine, valine, leucine, methionine, isoleucine, tryptophan, phenylalanine and histidine. In the horse, only lysine has been demonstrated to be essential. There is considerable interest currently in methionine and threonine but not enough information to advise on required dietary levels. The amino acids leucine, isoleucine and valine are of interest to those who work with high performance horses. The latter are the branched chain amino acids. Branched chain amino acids are broken down in the muscles during exercise and provide a source of energy. They also can use pyruvate generated during the breakdown of glycogen to regenerate alanine that is lost during exercise, resulting in a lower production of lactate. There is some evidence that BCAAs may also substitute for glycogen as an energy source during exercise. It is further theorized that by decreasing the concentration of tryptophan in the blood, BCAAs may help prevent the sensation of fatigue, although there are conflicting reports on this. BCAAs have been used to help prevent tying-up. Their use was associated with absence of symptoms and drop in blood levels of muscle enzymes with exercise. Most recipes use 30 grams of BCAAs and roughly equal amounts of valine and leucine, slightly less than half as much isoleucine. Valine also is reported to have a stimulant effect.

Supplementation

Supplementation may be helpful during aerobic exercise, prolonging the time to depletion of glycogen stores and sparing the BCAAs in the muscle from degredation (the muscle will take what it needs from blood if supplies are high enough). Some types of tying-up may benefit from treatment with BCAAs.

BCAAs should be administered approximately one half hour before and one half hour after exercise. For exercise bouts longer than two hours, repeat treatment during exercise may be indicated. Routine supplementation is not recommended.

Deficiencies

Deficiency of any of the essential amino acids will lead to poor growth and poor tolerance to stress. Specific symptoms for an isolated amino acid will depend upon those organ systems where it is particularly important. In the case of BCAAs, loss of muscle bulk and poor exercise tolerance may be seen.

Toxicity

None known.

Interactions

There are many complex interactions between amino acids and other amino acids. The proper functioning of amino acids is also closely tied to adequate energy content in the diet and normal vitamin and mineral levels. BCAAs restore levels of alanine and glutamine lost during exercise and have a sparing effect on muscle glycogen.

Indications

As above, BCAAs may be of benefit during aerobic exercise, particularly endurance events. Some horses with tying-up may also benefit by treatment with BCAAs.

CHAPTER 5

NUTRITION AND PERFORMANCE

With optimal nutrition, horses performing in endurance events may be safer and have better results.

ABOUT THIS CHAPTER

Nowhere are the effects of diet, for better or worse, more obvious than with a performance horse. Whether your demands are relatively modest or you are expecting the horse to exercise to the absolute limit of his capacities, gaps in the horse's nutrition will become evident very rapidly.

As already mentioned many times, the guidelines we have for feeding horses are based only upon estimates of what a horse needs to stay alive and reasonably healthy/normal when not being stressed. They often have little to do with any scientifically obtained information on actual needs but are rough estimates derived from knowing what types of food the average horse eats. This will be

adequate to let the horse get by if he is not doing much in terms of forced exercise but can be expected to fall apart under demands for heavy work. Horses can show, event, cut cattle, trail ride, race and compete in endurance events without ever getting supplements. The challenge is to optimize their nutrition so that they do it even better, faster, safer and easier.

This chapter looks at the basic diet for horses involved in various activities and explores how manipulation of diets and provision of supplements can improve performance. What is right for an endurance horse will not be right for a race horse. This chapter will point out those differences.

FUELS FOR EXERCISE

To perform, the horse's muscles must have fuel. The cycle of muscular contraction, relaxation and contraction again requires energy at each step. Without a sufficient amount of proper fuels to generate that energy, performance falls off. The muscle can obtain these fuels either from the blood or from energy stored inside and around the muscle cells. There are three basic types of fuel in the body—carbohydrates, fats and protein. We will look at each of these and how important they are to the process of muscular work.

CARBOHYDRATES

Carbohydrates are the premier fuel of muscle. Only carbohydrates can be used for both low intensity/endurance and high speed activity. The internal chemistry of the muscle of the horse (and all mammals) is set up to utilize carbohydrates. The other fuels can also be used but only if a critical amount of carbohydrate is also available. The idea that there is any other food/energy source that can substitute for carbohydrates is basically incorrect.

We are familiar with carbohydrates as sugars and starches. The horse also has these in his diet. In addition, horses are capable of digesting and using the carbohydrates present in tough plant materials (e.g., hays and grasses) that humans and creatures with a more simple intestinal tract cannot utilize. The basic equine diet is usually composed of 85 to 90 percent carbohydrate while the average human diet contains only about 30 to 45 percent. When you hear the term low carbohydrate applied to any diet that a horse would actually consume, realize it refers only to low sugar/starch levels, not total carbohydrate in the diet.

When the horse eats, the carbohydrates consumed are either used immediately as an energy source for all tissues of the body, stored in the form of a compound called glycogen inside the muscle cells or converted to body fat and some amino acids.

FATS

Fats comprise a very small portion of the natural equine diet, probably around three percent or less. Some fat ingestion is necessary to provide essential fatty acids that the horse cannot manufacture himself. When the horse puts on weight in the form of fat, it comes from fats that were manufactured in his body. When a horse burns fat during aerobic exercise, the fats used are in the form of fatty acids that come from the breakdown of body fat. With the long term feeding of a diet that is higher than normal in fats (usually 10 to 20 percent of calories eaten coming from fat either as added vegetable oils or specially processed animal fat/lard), there may be some stimulating of enzyme systems within the muscle cells that are involved in the use of fats as a fuel, leading to improved use of the fats. However, short term feeding of increased levels of fats has been clearly shown to have no effect on the fuels used during exercise. The muscle cell still prefers and uses the same amount of carbohydrate, its premium fuel.

There has been great interest lately in adding fat to horses' diets. Fat is inexpensive and is capable of generating more energy than an equivalent amount of carbohydrate during aerobic (low intensity/low speed) exercise. The burning of fats stored during low intensity/low speed exercise also allows the muscle to conserve its critical supplies of stored carbohydrate—i.e., glycogen stores. A certain amount of fat can be beneficial in these ways.

It is important to remember, however, that the horse's entire metabolism and especially the functioning of the muscles is geared to the use of carbohydrates. Even so called high fat diets for horses will only contain 10 to 20 percent of their calories as fats compared to the normal three percent. This would be considered an abnormally low fat diet for a human.

PROTEIN

The risks and benefits of protein to horses is an area wrapped in multiple layers of myth and misunderstanding. Adequate protein is absolutely essential to normal growth, health and athletic performance as well as the maintenance of healthy organs and tissues. The hoof and hair are well over 90 percent protein. Muscle cells contain huge amounts of protein. Protein is the material used to make DNA. The list goes on and on. Despite this, the myths remain that protein is somehow harmful and can cause problems such as bone and joint disease in growing horses or kidney problems in adult horses. Research in this area has proven these myths to be false but the opinion that horses need to be fed a restricted diet of protein persists.

Every performance horse, particularly high performance horses, should have a relatively lean body type with just enough fat as reserve to cover their ribs and well formed musculature. Hindquarters should be rounded and not "caved in" and all bones of the spine covered over with fat and muscle. Heavy racing schedules/endurance training and competition place extreme demands on the body. Simply providing enough calories to hold weight is a challenge. However, an extremely thin and "sucked up" appearance can have its root in insufficient quality protein as much as in insufficient total calories. Protein is broken down every time a horse exercises. Some of this can be captured and recycled within the body; some cannot.

DIFFERENT USE REQUIREMENTS

LIGHT USE

The horse will require from 1 to 1.5 percent of his body weight in total feed intake per day (1,000 pound horse needs 10 to 15 pounds of hay and grain daily). Ideal hay for a performance horse is 50:50 mixed high quality grass and alfalfa. Light use may require only hay. Individual animals that cannot maintain condition on hay alone may have grain substituted for a portion of the ration. Feed grain only on days the horse is worked. To begin with— one or two pounds two or three times a day. Minimum suggested protein level 12 percent.

MODERATE USE

Requires 1.5 to 1.75 pounds of feed per 100 pounds of body weight (1.5 to 1.75 percent of body weight) per day. Many horses in moderate use can maintain weight on hay alone, others cannot. Withhold or drastically cut (by half at least) grain on days the horse does not work. Minimum suggested protein level 14 percent.

ENDURANCE

Requires up to 2 percent of body weight (2 pounds per 100 pounds of body weight) per day. Endurance horses cannot maintain weight on hay alone. They simply physically cannot eat enough of it. Grain may need to comprise as much as half, even more, of the total daily intake of feed by weight.

Alternatively, calorie intake can also be boosted by the use of vegetable oil or processed lard. Long term health risks of using animal fat are not known and vegetable oil is a healthier source. Oils and fats provide three times the calories as grain on a weight basis. Advisability of feeding increased protein is a hotly debated topic in endurance horses. Many nutritionists fear increased protein intake will lead to increased ammonia and other waste products which could adversely affect performance. However, the ammonia generated during exercise is coming from breakdown of the horse's muscles and high energy ATP stores, not from the diet. Feeding 100 percent alfalfa hay will result in a total protein content for the entire diet of about 12 to 13 percent. This is a good start. Horses that are losing muscle mass despite this diet need more protein and calories.

RACING AND OTHER HEAVY USES

Heavy racing schedules severely drain body reserves. However, work between races, especially for Thoroughbreds, is often very light. Racing horses require 1.5 to 2 percent of body weight in total feed per day. Horses on very high grain intakes may be fed 100 percent alfalfa. Otherwise, mixed alfalfa and grass hay is best. Minimum suggested protein level is 14 to 16 percent.

SUPPLEMENTS AND SPECIAL DIETS
FOR PERFORMANCE

Supplements may help deliver improved performance by providing the nutriments a horse requires for a particular sport. A discussion of the nutritional needs of different use categories follows. Specific supplements are then presented individually. Not every horse will respond dramatically to a recommended supplement. There will be individual differences in which nutrients work best.

The Supplements for Performance Enhancement by Usage chart (page 149) summarizes recommended supplements by use, since different sports and uses place unique demands on a horse's metabolism (see this chapter's breakdown of supplements and Chapter 3, Nutrition A to Z for more detail).

Racing (and other high speed events)

The need for speed places unique demands on the horse's metabolism. There is only one source of fuel—carbohydrates—that can be metabolized quickly enough to generate the energy needed for speed. The horse also relies upon his stores of high energy compounds ATP and creatine phosphate to provide bursts of speed.

BASIC DIET—Your base of operations is in the horse's daily diet. Use the best quality hay you can find, preferably alfalfa/grass mix or straight alfalfa. With straight alfalfa, you need to supplement the horse's magnesium (see A to Z). High quality grain is also essential. Grain is your horse's source of carbohydrate, the speed and energy fuel.

Horses in training should actually gain weight. They are burning excess body fat and putting on muscle. Muscle is heavier than fat. Although a horse may look thinner or more streamlined, they should actually weigh more. Weight loss is disastrous to performance at any stage. You want to avoid that severely tucked-up appearance so common in heavily trained and raced horses. Many horses will hold their weight and muscle well on the above diet if they are fed enough. If the horse is not putting on muscle well, consider a protein supplement (see protein).

PRERACING—Everyone wants to find the magic

prerace treatment that will give their horse an edge. Unfortunately, there is no substitute for sound daily nutrition and an adequate training program.

There are no miracle pills or shots—at least none you will get away with for very long. The marketplace is flooded with prerace powders and pastes, all claiming to deliver improved performance. Some of these are helpful, some are nonsense. Not even the best supplement will help a horse that is basically unsound in legs or wind.

Endurance

BASIC DIET—The basic principles for the race horse's diet—need to maintain weight and muscle mass—apply as well to endurance horses. Endurance horses often have total caloric needs that exceed those of a race horse, based on the sheer amount of work they must perform. However, differences in the type of substrate they use to generate energy result in some important differences.

Studies of the muscles of endurance horses before and after competition clearly show the vast majority of work is performed aerobically. This means fat becomes a much more important energy source. Studies have confirmed that endurance horses use considerable amounts of fat during a race. The intramuscular stores of glycogen in the active, aerobic muscle fibers are depleted or nearly depleted by the end of an endurance race.

Simply maintaining weight on heavy training and competition schedules is a very difficult task for the endurance horse. Grain is a necessity; the horse simply cannot eat enough hay to provide the calories he needs. Limited time for eating is also a problem. Use of beet pulp or commercial concentrates containing beet pulp is a good choice for endurance horses. It is more calorie dense than hay but is digested in the hindgut, avoiding large fluctuations in blood sugar and providing a steady source of calories during work but with less bulk/ weight than hay. Addition of fat to the diet is also common with endurance horses as a concentrated source of calories.

However, if the muscle had a preference it would use carbohydrate. The muscle is also obligated to use a certain amount of carbohydrate to maintain all the intermediates of the Krebs cycle (or citric acid cycle)—the energy pathway that burns fuel aerobically. Carbohydrates will also be needed to provide the energy for any bursts during a race and for the final kick. If you fail to pay attention to the endurance horse's carbohydrate needs, you will get a flat performance and early fatigue.

Horses will manufacture the glucose they need to maintain blood levels from both fat and the volatile fatty acids produced in the large intestine when they break down hay. This will maintain muscle glycogen stores at a minimal level but will not give the horse any reserve, or any advantage. To best handle the demands of an endurance race, the horse needs to have ample stores of glycogen in his muscles (see carbohydrate loading). Carbohydrate loading is well documented to benefit human endurance runners. It is so successful, it is almost a universal procedure. While diets and digestive tracts of people and horses differ, the way their muscles use fuel does not.

The Three-Day Event Horse
The three-day event horse is listed separately because the energy requirements for this unique sport

fall in a zone between those of endurance and those of speed. This is largely due to the extreme demands of day two.

A diet of adequate energy with high quality protein is as essential for this horse as the above two. Studies of muscle biopsies from three-day horses have clearly shown they rely much more heavily on anaerobic metabolism on day two than endurance horses do during a race. This is probably partially a function of less endurance type training and partially because the speeds required exceed those for endurance horses. Carbohydrate loading is of clear potential benefit for three-day horses. The addition of moderate amounts of fats, such as vegetable oil, may also be of benefit (4 to 6 ounces a day).

If you are currently competing at the Horse Trial level, follow recommendations for low and other moderate use horses.

Other Low and Moderate Use Activities
The vast majority of horses competing in other equine activities or used for pleasure purposes will handle this activity very well if you pay careful attention to providing an adequate, high quality basic diet and supplementation as detailed for all performance horses.

LEGEND

Use Icons	Probability of Deficiency					Severity of Deficiency & Risk of Toxicity
Maintenance	None	Little	Moderate	High	Heavy	None
Light	None	Little	Moderate	High	Heavy	Little
Moderate	None	Little	Moderate	High	Heavy	Moderate
High	None	Little	Moderate	High	Heavy	Serious
Heavy	None	Little	Moderate	High	Heavy	

SUPPLEMENTS FOR PERFORMANCE ENHANCEMENT

Usage	Lipoic Acid	B Vitamins	Carnitine	BCAAs	Creatine Loading	Bicarbonate, Citrate & Phosphate Loading	Fat	Carbohydrate Loading	Protein	Electrolytes	Chromium	Grapeseed	Selenium	Vitamin E	Bioflavinoids	Vitamin C	DMG
												Antioxidants					
Racing (and other high speed)		X	X	X	X	X*	X	X	X	X	X	X	X	X	X	X	X
Endurance		X	X	X	X		X	X	X	X	X	X	X	X	X	X	X
3-Day Eventing		X	X	X	X		X	X**		X	X	X	X	X	X	X	X
Low/Moderate Competition		X	X								X	X	X	X	X	X	X

* Illegal in Standardbreds
**Upper level

Nutrition and Performance   149

B VITAMINS

Probability of Benefit	Dosage	Toxicity	Complementary Element	Usage

	Biotin – 0-20 mg
	Cyanocobalamin-
	  Vitamin B12 – 0
	Folic acid – 20
	Niacin – 50-100
	Pantothenic acid –
	  100-200
	Pyridoxine – 120-300
	Riboflavin – 20 mg
	Thiamine – 250 mg
		Small to none. Thiamine in doses of over 500 mg may produce some sedating/tranquilizing effects, with or without effect on performance.	B vitamins work best as a group.	

Indications / Rationale

The B vitamins are essential to the use of all types of fuel/food, especially carbohydrates. Stress of any type, including competition, decreased appetite, digestive problems and periods of fasting all increase the need for B vitamins.

Method / Timing

In general grains, bran and yeast provide all the B vitamins required in a horse's diet. However, there are exceptions. B vitamins are indicated when a horse is not eating normally (or at all), in horses that are heavily exercised or stressed for any reason (age, surgery, injury, infection, shipping, etc). Horses with poor quality hooves may benefit from biotin supplementation of 10 to 20 mg a day, with higher doses indicated for horses receiving little or no grain. B12 supplementation is not a consideration except in horses with prolonged periods off feed with intravenous fluid nutrition (such as following surgery). Horses being heavily exercised or denied access to fresh grass may benefit from folic acid supplementation. High grain intake may call for supplemental thiamine. Thiamine is also used in higher doses (500+ mg per day) to prevent tying-up. Horses that are jumpy, nervous and difficult to work around may manifest symptoms of B vitamin inadequacy. See Chapter 3, Nutrition A to Z, for dosages adjusted by diet type.

BICARBONATE

Probability of Benefit	Dosage	Toxicity	Complementary Element	Usage
	The effective dose is 250 to 350 mg/kg or 125 grams for a 500 kg (1,000 pound) horse two hours out. That's a lot of bicarbonate! It adds up to 21 to 25 tablespoons of baking soda. Most horses will not voluntarily consume this much and it must be given as a paste or by stomach tube.	Bicarbonate loading is illegal in Stan–dardbreds (which is a testimonial to how effective it is).	None	

Indications / Rationale

When a horse works at speed, the interior of the muscle cell becomes acidic. After a certain level of acidity, energy generating pathways may fail to work efficiently. However, the most recent studies have shown giving bicarbonate does not change the pH inside muscle cells, only in the blood, so the exact way in which bicarbonate helps is unknown. When a horse is given bicarbonate before a race, the pH of the blood rises (low pH-acid). This encourages acidity to leave the cell, and it is neutralized in the blood and other tissues.

Method / Timing

Given as a paste or through a stomach tube. Given before a race. Most effective for horses two hours out.

BIOFLAVINOIDS

Probability of Benefit	Dosage	Toxicity	Complementary Element	Usage
	16-40 grams a day	No known toxicities are associated with the use of naturally occurring mixtures of bioflavinoids (i.e., citrus extracts).	Vitamin C	

Indications / Rationale

Bioflavinoids are a class of vitamin-like substances which enhance the effectiveness of vitamin C. They have various biological effects, including the strengthening of capillaries and anti-inflammatory action. They are plentiful in a wide variety of plants, especially those that contain signficiant amounts of vitamin C.

In general supplementation with bioflavinoids can be expected to have the same good effects as with vitamin C. Horses can benefit from supplementation with bioflavinoids in terms of fewer infections, better wound healing, better stress tolerance and improved ability to detoxify drugs and pollutants.

Lung bleeding in many horses responds well to supplementation with bioflavinoids (hesperidin complex), vitamin C and grapeseed extract.

Method / Timing

Feed on a daily basis.

BRANCHED CHAIN AMINO ACIDS

Probability of Benefit	Dosage	Toxicity	Complementary Element	Usage
	30 grams total per day of approximately equal amounts of valine and leucine, half as much isoleucine. For race horses, ideally about an hour before work/ race and during first hour after racing.	Moderate with racing animals.	Vitamin B6 Carnitine	

Indications / Rationale

BCAAs could contribute greatly to the performance of endurance horses. Muscle breakdown and muscle wasting is a significant problem with endurance horses. It may be largely prevented by prerace use of supplemental BCAAs and by repeated administration at timed intervals (approximately every two hours) during the race. Lactate accumulation and metabolic acidosis are consistent findings in endurance horses at the end of a race. BCAA supplementation may favorably modify this problem. BCAAs also have somewhat of a glycogen sparing effect. BCAAs (valine, leucine, isoleucine) have definite value in improving recovery from a race, maintaining muscle mass and helping prevent postrace muscle soreness. They are advertised as a performance aid and treatment for horses with classic problems of tying-up or persistently elevated muscle enzymes.

Studies in horses have shown that pre-exercise use of BCAAs results in reduced amounts of free lactic acid in the blood and significant decreases in blood enzymes that signal muscle damage. Extensive breakdown of muscle stores of BCAAs occurs with each bout of exercise. If you supply them in the correct amount and at the correct time, the muscles will take up the increased BCAAs in the blood and use them instead, preserving muscle supplies.

The muscle can also take the breakdown products of BCAA use and combine these with the lactate produced to make alanine, another amino acid that is used by the muscle in large quantities during exercise. Finally, there is evidence to suggest that supplying extra BCAAs before exercise may help delay the onset of fatigue by preventing changes in brain chemistry that are believed to trigger the feeling of fatigue.

As an alternative to use of BCAAs, HMB, which is a metabolite of the BCAA leucine, can be used for similar effects and benefits.

Method / Timing

Thirty grams given within one hour of exercise and every two hours during prolonged exercise. My advice would be to try it out first before using BCAAs prerace. Pick a day when the horse will be having a stiff training session and give the BCAAs at the same time pretraining as you would prerace. If you like the response, use them.

If you don't, it might be best to save BCAAs for use immediately after the race. There is no question about their benefits then.

CARBOHYDRATE LOADING

Probability of Benefit	Dosage	Toxicity	Complementary Element	Usage
	There are various ways to load carbohydrates.		Chromium Water B Vitamins	

Indications / Rationale

Carbohydrate loading is a process developed for human weight lifters and runners to increase their supply of muscle glycogen—the storage form of carbohydrate inside muscle cells. It is of proven benefit to human athletes and over the past few years has also proven itself with horses. Simply put, the horse is fed an increased level of readily digestible carbohydrate for three to four days preceding an event.

At its simplest carbohydrate loading can be accomplished to some degree by increasing the amount of grain you feed by 10 to 20 percent. Some people cut hay back by an equivalent amount, others recommend always having ample hay available. Grain consumed after the horse is worked will be the most effective meal in terms of glycogen/carbohydrate loading.

Method / Timing

Method: There are also various commercial carbohydrate powders available. These are added to the grain. Manufacturer's directions range from no guidelines at all on some products to amounts of half to one cup of carbohydrate powder per feeding, two to three times a day.

Use the lower end for small horses, upper end for larger animals. Choose a product that has a blend of sugars. You can get a good boost in blood sugar to trigger insulin release (the insulin drives the sugar from the blood into the muscle cells). Products with sugars whose chemistry is all very short chain lengths (like table sugar) cause a rapid rise and rapid drop in glucose. This is only useful in the one to two hours after the horse is worked.

For other times, when muscle is not as immediately hungry for carbohydrates, longer chain molecules work better since they give a more gradual rise and drop, allowing more time for the sugar to be taken up. The label should state the product contains either malto dextrans from a natural source, such as corn, or give the specific chain lengths the product contains.

Critics of carbohydrate loading in endurance horses use many arguments. One is that carbohydrate feeding generates more body heat. This is true (although not to the extent you would think). However, carbohydrate loading is largely completed before the race and need not

involve a big grain meal the day of the race.

You may hear many cautions about carbohydrate loading and inducing laminitis or endotoxemia. This might be a concern with feeding the horse an extra 10 or 15 pounds of corn or sweet feed but efficient carbohydrate loading involves the use of specially designed carbohydrate powders which are very highly digestible and probably do not even reach the large intestine in any significant amount.

Amounts used are equivalent to, or less than, those used when doing glucose tolerance tests and absorption studies in horses and in my opinion are safe to use in normal horses.

You also hear that fat is a superior fuel for endurance horses, providing far more calories than carbohydrate per gram and burning with less generation of heat. This is true but has little to nothing to do with muscle metabolism during an endurance race. Endurance horses mobilize fat from body stores and intramuscular storage during a race and burn this very efficiently. The mobilization is triggered by hormonal and biochemical changes in the muscle and blood—one of the most important of which is depletion of glycogen stores. The muscle will always opt to use carbohydrate first, unless work is of a very low intensity.

It is also interesting to note that recommendations

made for supplementing horses during an endurance race always include carbohydrate rather than fat. A typical recommendation found in product literature calls for five liters of water, 30 grams of sodium chloride (or 60 grams of a mixed electrolyte preparation) and 15 grams of sucrose or glucose every two hours. No mention is made of fat. This is because the fat used by the muscle during exercise comes from body stores, not diet.

Timing: For race horses, feed the extra carbohydrate 2-3 times a day for 3-day prerace. May also try just feeding it routinely, on days the horse works, within one to two hours of feeding. For one of the feedings, try to feed the horse as soon as he is cooled out—within one or two hours of working—to get the maximum uptake by the muscle.

CARNITINE

Probability of Benefit	Dosage	Toxicity	Complementary Element	Usage
Possible but unclear, more likely to occur with long term daily use than pre-event use.	6 to 12 grams per day		None	

Indications / Rationale

Has been suggested to improve endurance by enhancing or facilitating the use of fatty acids as a muscle fuel. Encourages continued aerobic energy generation by handling waste products (acyl groups) generated during entry of pyruvate into oxidative cycle.

Research in other species concerning carnitine and athletic performance is very equivocal. Some studies claim a benefit for both endurance and speed events while others find absolutely no effect at all. In one study, feeding 5 grams per day resulted in decreases in lactate production and muscle enzyme elevations during and after exercise, although improved performance per se was not proven.

Carnitine is at least of theoretical benefit to racing horses as it may help buffer the byproducts of using pyruvate, a step in the anaerobic metabolism of carbohydrates.

It serves the same function in the oxidization of fats (also aerobic) and is essential in getting the high energy free fatty acids into the mitochondria, the furnaces of the cell where oxidative burning of carbohydrates and fats occurs.

Another problem with carnitine supplementation is that all the studies in other species that found a benefit involved long term supplementation with carnitine, for several weeks, while equine products including it are usually used only right before the race.

You need to focus on increasing the efficiency of anaerobic metabolism to get peak production of raw speed and carnitine is primarily of use in aerobic pathways. There is simply no evidence that giving carnitine will make the horse go faster.

It would theoretically be of more potential benefit to Arabians or Thoroughbreds, which have a high aerobic capacity than to Standardbreds or quarter horses. Carnitine is also of more theoretical benefit to endurance horses than horses in short or more intense events. This is particularly true in light of evidence in other species that carnitine levels may become abnormally low during long term events, which would effectively cripple the muscle's ability to effectively use fats as an energy source.

Method / Timing

If you want to try carnitine, I would suggest doing so before an actual race, because of carnitine's stronger association with aerobic than anaerobic metabolism. However, the value of a single prerace dose of carnitine is highly questionable. Most will probably be lost in the urine and never make it to the muscle cell before the time in the race when the horse needs it and might take advantage of increased blood levels. Either plan to use

the supplement at regular intervals throughout the entire endurance race or try a program of longer term daily supplementation.

Begin supplementing with five to six grams a day late in the horse's training when you are putting on the finishing touches. Allow about two weeks or so before expecting improvements.

CHOLINE

Probability of Benefit	Dosage	Toxicity	Complementary Element	Usage
Possible but unclear.	Not well established, but probably around 10 grams daily		None	

Indications / Rationale

Has been suggested as an aid to prevent depletion of acetylcholine (a chemical released from nerves during muscle contraction) during endurance exercise and to enhance acetylcholine production in general, leading to improved muscular performance.

Method / Timing

Once daily for three days before competition.

CHROMIUM

Probability of Benefit	Dosage	Toxicity	Complementary Element	Usage
	1,200 to 1,600 mcg per day	Unknown, do not exceed recommended dose.	Zinc Carbohydrates	

Indications / Rationale

Benefits of using chromium include enhanced uptake of glucose by fat and muscle tissues, enhanced liberation of fatty acids from fat stores during aerobic exercise and stabilization of blood glucose levels which avoids extreme peaks and valleys often seen after a heavy grain meal.

Chromium is a mineral present in trace amounts in the body, which is critical to proper functioning of insulin. The addition of chromium is suitable for all types of performance horses.

Chromium improves the effectiveness of insulin, leading to improved uptake of glucose by muscle cells and enhanced release of fatty acids for energy during aerobic work levels.

I have encountered several performance horses with extremely low blood sugars between grain feedings (e.g., morning prefeeding blood sugar), elevated resting and post feeding insulin levels with blunted blood glucose response to feeding, all consistent with over secretion of insulin and insulin resistance. Chromium normalizes blood sugar values within a very short time and leads to improved alertness and performance.

Method / Timing

Feed daily.

CITRATE

Probability of Benefit	Dosage	Toxicity	Complementary Element	Usage
	It takes 220 grams of potassium citrate to get the same effect as full bicarbonate loading with baking soda. Using this much is also likely to get you a milkshaking positive.		None	

Indications / Rationale

Citrate is used as an alternative formula to bicarbonate. It's really just a bicarbonate precursor and is turned into bicarbonate by the body. It takes 3.5 grams of potassium citrate to equal just one teaspoon of baking soda.

Method / Timing

Given as a paste or through a stomach tube. Given before a race. Most effective for horses two hours out.

CREATINE

Probability of Benefit	Dosage	Toxicity	Complementary Element	Usage
Unclear	See Method/Timing	Minimal. Some muscle and intestinal discomfort is possible in sensitive individuals on a high dose.	Carbohydrate loading DMG	

Indications / Rationale

Creatine is the only form of energy store that can provide the horse with peak powers of acceleration. Creatine probably has little to offer the endurance horse in terms of overall performance, but the nutrient is of considerable potential benefit to race horses.

Creatine is the base of the high energy compound creatine phosphate (CrP). This is a stored form of energy, which is instantly available to the muscle cell for rapid acceleration. When a horse leaves the gate or makes a big move during a race, he is calling upon his muscles to instantly provide more energy. Generation of energy from fat or carbohydrates takes too long for this kind of quick kick. The horse will need to rely on energy stores in the form of ATP and creatine phosphate. The more on hand, the more ability for acceleration.

There is no known way to improve upon a horse's level of ATP stores. The muscle has a set amount and must begin to replenish it by burning carbohydrates or fats as it is used up. When ATP releases its energy, it loses a P (a phosphorous). The role of creatine is to instantly replace that phosphorous molecule from its own supply, changing from CrP to free Cr in the process. Work

in lab animals, humans and to a limited extent in horses has shown that increased intake of creatine in the diet can cause an increase in the muscle stores of both free Cr and the high energy CrP form.

Creatine is naturally found only in meats. Raw meat has the highest concentration. The body can manufacture creatine from the amino acids alanine and glycine. Athletes who are vegetarians have lower muscle stores of Cr and CrP than meat-eating athletes. Supplementing their diet with creatine raises their creatine and CrP levels and improves performance. The same happens with lab animals. Since horses are vegetarians, the theoretical potential for improvement in speed is there.

With increased creatine stores, the horse will not actually have more speed but will be able to hold that top speed longer. If your horse comes off the gate well but quickly loses ground to the speedballs, you should be able to hang in there longer and get a better position. If you know from experience you can only use the horse hard once in a race, you may find you can now use him twice or his one move is now one BIG move.

Method / Timing

The author has had some success with creatine in several horses by either giving it in a very small amount of grain with about 100 grams of carbohydrate powder or by stomach tube. Results have been obtained using a specially stabilized form of liquid creatine in a honey base as well as powdered.

If using this large amount, follow a loading schedule of creatine administration for three days before the race. However, some people have reported cramping with creatine loading. To be on the safe side, begin your load-

ing five days out from the race. On day four and race day, give only a maintenance amount of creatine of 12 grams. An alternative is to give the horse the maintenance dose of creatine of 12 grams per day on the day that he works. Within about two weeks (longer if the work schedule is light), creatine levels in the muscle should start to rise.

This is a reasonable plan to follow with horses that are in late training, almost ready to race. It gives you a chance to get some idea of exactly what the creatine can do for you with a minimum of complicating factors. The horse will be pretty much as fit as he is going to be when he starts racing, eliminating any conditioning effects on performance, and by evaluating the horse in training sessions you will not have the confusion of an actual race making it difficult to say if the horse gave all he had.

However, creatine is highly unstable. Moisture causes rapid degradation to the point that it is virtually useless within 15 minutes. It is not something you can have mixed into a batch of feed at the feed mill. It should also not be mixed with a large grain meal since the horse's saliva and moisture in the feed (e.g., from molasses) starts degrading the creatine on contact. The creatine needs to make its way out of the stomach and into the small intestine rapidly so it can be absorbed. This requires an empty stomach. Uptake of creatine by the muscles is greatly enhanced by increased blood glucose and an insulin release. You therefore also should deliver a substance that will cause this reaction at the same time as the creatine.

Failure to follow all of these guidelines will result in less than optimal results and probably explains why many creatine trials have failed. Researchers have also noted that the muscle has an upper limit for creatine and creatine phosphate that is probably genetically determined. Once a muscle is at this level, it is impossible to get any more creatine in regardless of how much the individual consumes. Although studies do not yet exist to prove it, there is a good chance that horses which are already elite performers have high natural levels of creatine and creatine phosphate that cannot be changed by supplementation.

DMG-DIMETHYLGLYCINE

Probability of Benefit	Dosage	Toxicity	Complementary Element	Usage
High in some circumstances.	1,500 to 3,000 mg per day		Carnitine Lipoic acid B vitamins CoQ$_{10}$	

Indications / Rationale

Dimethylglycine is a naturally occurring (found in the body) chemical that is used to enhance muscle metabolism. The product first emerged for human athletes after a rather suspicious piece of research reported it greatly enhanced aerobic metabolism. Despite multiple human studies since then, no benefit has been found in athletes. However, testing of a related compound, trimethylglycine (TMG), in horses did show some benefit for animals that are in the early stages of training (not fit). The benefit disappears as the horse becomes conditioned. Whether there is additional, ongoing benefit for horses with histories of muscle problems, such as tying-up, remains to be proven, although it is widely used for this purpose. DMG or TMG are also sometimes used in the treatment of arthritis, for their antioxidant properties.

Method / Timing

Feed daily, 3,000 mg for first two weeks, then 1,500 mg/day.

ELECTROLYTES

Probability of Benefit	Dosage	Toxicity	Complementary Element	Usage
	Varies	High (dehydration if fresh water is not always available).	Horses that are prone to tying-up or muscle soreness may benefit from supplementation with additional magnesium, potassium and water.	

Indications / Rationale

Hay and salt blocks fulfill electrolyte needs of most horses. Provisions of supplementary electrolytes becomes important for all performing horses under conditions of either heavy prolonged work and/or work in the high heat with significant sweating.

The horse's requirement for supplemental electrolytes relates directly to how much work he performs in the heat. If the horse takes in adequate water and food and has constant access to a salt block, no specific supplementation will be needed under most circumstances. However, it is a good idea to keep a close check on how well the horse is hydrated using the pinch test. Pick up a fold of skin on the horse's neck between your fingers. The skin should spring back into place quickly, not remain tented up. If it does not, the horse is dehydrated because of insufficient salt intake, insufficient water intake or both. Providing supplemental electrolytes in the feed or in water, with a separate untreated water source may correct the problem.

Horses on Lasix are a special case and will benefit from measures taken to counteract the electrolyte and fluid losses caused by Lasix. Lasix does NOT work by causing a fluid loss and drop in blood pressure from dehydration. This is only a negative side effect. Lasix has a direct and specific effect on how well the right side of the heart—the side that pumps through the lungs—operates. This is a chemical effect, not one related to how much the horse urinates.

For example, other diuretics do not give the same benefits as Lasix because they do not have the same effect on the heart. It is also possible to completely eliminate the benefits of Lasix to the heart by giving the horse phenylbuzatone even though the horse produces just as much urine. Horses that do not improve as anticipated on Lasix are usually either getting too much or having too much water withheld along with giving the Lasix.

Have you ever noticed how some horses on Lasix shake after they are worked? This is a direct result of dehydration and electrolyte depletion caused by Lasix. Lasix causes a massive immediate loss of potassium, an electrolyte critical to normal and efficient muscular functioning. It also causes lesser losses of calcium and magnesium, the other two electrolytes critical to muscle. Add to this an element of dehydration and the horse will definitely not perform well.

To retain the benefits and eliminate the side effects of Lasix, a race horse should be supplemented with potassium chloride and magnesium sulfate (2 to 4 grams of each per day, depending on laboratory analysis of the blood levels) daily for three days before the race. It is also my usual practice to tube the Lasix horse with about six quarts of water either the night before or about eight hours out from a race. Some of you may remain convinced that this is a ridiculous thing to do—until your Lasix horse is performing below your expectations and you finally try it.

Method / Timing

Endurance horses have unique electrolyte requirements that are largely related to the inevitable large production of sweat during prolonged exercise of this type. There are also losses arising from muscle in prolonged exercise of this type. Endurance horses show profound and persistent deficiencies in calcium, bicarbonate, chloride, potassium and sometimes magnesium after an endurance ride. Magnesium losses/needs are probably underestimated since the blood level of this mineral does not accurately reflect muscle levels and reports of some low or even borderline magnesium levels after an endurance event probably means we are seeing only the tip of the iceberg.

Electrolyte supplements for both daily use and use during a race should include all of these minerals in generous amounts. Look for one that contains potassium chloride as well as magnesium and calcium salts in ratio of at least 1:2 to sodium chloride. Bicarbonate or citrate will have to be added separately.

Interestingly, studies on 3-day horses show patterns of electrolyte depletion almost identical to those of endurance horses, largely precipitated by sweat losses. Follow the same guideline as those provided for endurance horses.

Most electrolyte mixtures are predominantly salt (sodium chloride) as they should be. These are fed at a rate of one to three ounces a day depending on heat and level of exercise, as well as at intervals during endurance events.

Follow guidelines above for choosing supplements for endurance or Lasix horses, which have special needs.

For routine daily use, add to feed or a SEPARATE bucket of water. ALWAYS provide fresh unsupplemented water as well.

FAT

Probability of Benefit	Dosage	Toxicity	Complementary Element	Usage
High, but controversial.	See Method/Timing		Carnitine B vitamins Inositol Lipoic acid	

Indications / Rationale

Fat has a far more important role to play in the diet of the endurance horse than perhaps any other discipline. Feeding additional fat on a regular basis (but not with the total exclusion of grain) is beneficial in that the horse will be guaranteed an adequate supply of fat for energy and will be less dependent on intramuscular glycogen stores. This is the important glycogen sparing effect you may have heard mentioned. Feeding fat may also train the muscle to use fat more. Horses fed a 45:45:10 hay:grain:fat diet can adapt to increased levels of fat and show less breakdown of glycogen with work, which will leave them more reserve at the middle and end of the race. The addition of fat also cuts down on the total volume of feed the horse must consume, lightening the load during a race.

The controversy over fat and carbohydrate will probably continue for a long time to come. The ultimate answer may be a compromise between the two differences of opinion with carbohydrate fans conceding a role for increased fat in the diet and fat fans appreciating the indispensable contribution of carbohydrates.

Method / Timing

For weight gain purposes, liquid vegetable fats or processed animal fats (any fat supplement that is a solid/crystal/powder is animal fat-lard) are often used at a rate of anywhere from several ounces to 2 cups or more daily.

Fats are a common component of the diet for endurance horses, who have difficulty taking in enough hay and grain to hold their body condition during heavy training and competition. Fats may be used in the diet of other high performance horses, as well. However, feeding fat at high rates (over 10 percent of total calories consumed) can result in the enzyme generating systems inside the muscle cell changing toward a profile that favors the use of fats for fuel.

Fat fed immediately before work or during work cannot be used by the muscle. An exception to this might be MCTs—medium chain triglycerides—available in pure form at health food stores.

Substitution rate: 1 lb. grain = approx. 5.3 oz. fat.

GRAPESEED EXTRACT

Probability of Benefit	Dosage	Toxicity	Complementary Element	Usage
	0.6 to 1 mg/lb or as directed*		Bioflavinoids Vitamin C Vitamin E Selenium Zinc Copper Lipoic acid CoQ_{10}	

Indications / Rationale

Grape/grape seed extract contains many active antioxidant chemicals, including bioflavinoids and polyphenols such as proanthocyanidin. The latter in particular have extremely potent antioxidant activity, being from 20 to 50 times more active than vitamin C and vitamin E. Grape seed extract also assists the entry of vitamin C into cells. It helps to stabilize histamine release, making it a natural approach to allergy control. Grape seed extract protects small blood vessels from damage to their walls. The anti-inflammatory/antioxidant effects have led to its use in arthritis control and in preventing damage to heavily exercised muscles. General anti-aging benefits have also been proposed.

Helpful in controlling lung bleeding, tendonitis, myositis, periodontal disease, chronic inflammatory conditions such as recurrent uveitis (moonblindness). Grape seed extract topical poultices have been used to shrink melanomas—pigment containing tumors common to older grey horses.

Method / Timing

*Usual dosage is 0.6 to 1.0 mg/lb. Higher doses may be used for horses with specific problems that may benefit (e.g., lung bleeding, COPD/heaves, heavy exercise, etc.). Feed daily.

INOSINE (AKA RIBOSE, HYPOXANTHINE RIBOSIDE)

Probability of Benefit	Dosage	Toxicity	Complementary Element	Usage
	18 to 30 grams per day or one hour before the event (no established equine doses)	Has been shown to decrease performance in humans	None	

Indications / Rationale

Has been suggested as an aid to reformation of ATP stores during exercise but may have adverse effects on performance. Inosine, which may also be listed as hypoxanthine riboside on a label) must be mentioned here as it is marketed as an endurance performance enhancer. With the single exception of an obscure report out of Russia, there is absolutely no evidence to support the use of inosine in any species, including the horse. In fact, studies of human endurance athletes have more than once shown inosine adversely affects performance.

Inosine is of value only in a few, rare, genetically inherited disease states or under other extremely unnatural experimental conditions (e.g., depriving an isolated heart completely of oxygen for prolonged periods of time). It is a waste product, produced when the body completely metabolizes its stores of the energy compound ATP. There is no evidence at all that providing inosine during exercise will stimulate the muscle to manufacture ATP.

Method / Timing

There is no evidence at all that providing inosine during exercise will stimulate the muscle to manufacture more ATP. Unless more evidence surfaces to support its use, this is one supplement you should not even bother to experiment with.

L-ARGININE

Probability of Benefit	Dosage	Toxicity	Complementary Element	Usage
Unknown	12 g/100 kg		B-6 Gamma oryzanol	

Indications / Rationale

L-arginine is used by human athletes to trigger the release of growth hormone from the brain. Growth hormone is a powerful stimulator of muscle, resulting in increased mass and strength with decreased body fat—in short, it is an anabolic. Release of growth hormone is triggered by exercise and is largely responsible for the improved muscle mass and strength resulting from training. Arginine has been proven to enhance growth hormone release in human athletes. Studies on horses are lacking.

Use of gamma oryzanol at least theoretically would complement L-arginine. However, the effectiveness and safety of this combination is unknown. The amino acids L-ornithine alpha-ketoglutarate and glycine have also been used in humans for a growth hormone releasing effect. Equivalent equine doses would be in the range of 40 gm/day of glycine and 6 gm/100 kg/day of L-ornithine alpha-ketoglutarate. Again, the effectiveness and safety of trying this in horses is unknown at this time.

Side effects of headache, nausea and diarrhea have been reported in humans and are certainly theoretically possible with horses. Any horse showing signs of abdominal pain, depression, irritability or any other type of adverse reaction after dosing should not be given L-arginine. Use in horses is largely uncharted territory and you will be proceeding at your own risk if you try this supplement. L-arginine can also cause activation of latent virus in the body in humans (e.g., herpes outbreak - cold sores or genital). Long term safety has not been established in any species.

Method / Timing

Supplementation with arginine to get the growth hormone releasing effect is tricky. It must be given as the pure amino acid and recommendations suggest giving on a completely empty stomach and several hours distant from the intake of any other protein source. Some human athletes get up in the middle of the night to take their arginine. L-arginine should NOT be given on days the horse does not exercise and should be used in cycles of no more than 12 weeks on the supplement with 6 to 8 weeks off. L-arginine tastes terrible. You will have to dissolve it in a small amount of water and administer by dose syringe. There are L-arginine products for horses on the market. Recommended dosage is usually 30 grams/day. However, effective dose in people is 12 gm/100 kg which would translate into 60 gm/day for a 500 kg (1,100 pound) horse.

LIPOIC ACID

Probability of Benefit	Dosage	Toxicity	Complementary Element	Usage
	300-1,200 mg or more		Antioxidants Carnitine	

Indications / Rationale

Lipoic acid is a potent antioxidant that is active in all body tissues, both inside and outside cells. It is also essential to the efficient generation of energy.

Method / Timing

Lipoic acid can be found in supplements manufactured for high performance horses and/or horses with muscle problems. To give lipoic acid as a trial, realize that it will take time to work. Giving it on a one-time-only basis a few hours before a performance will not be as effective as using it for a few days beforehand. It may take weeks of supplementation to establish benefits. If you do plan to use it for longer than a few days at a time, stay with the lower doses.

PROTEIN

Probability of Benefit	Dosage	Toxicity	Complementary Element	Usage
	Most supplements manufactured for horses are about half soybean meal and half milk protein. Half to one pound daily of such a supplement will be needed.	Digestive upset is possible.	B-6	

Indications / Rationale

If your horse is not putting on muscle well, consider a protein supplement. If digestive upset develops (gas, loose manure, bloating, etc.) while using a supplement made of half soybean meal and half milk protein, it is more than likely that the soybean is to blame. You may then need to go to your health food store and shop for a protein powder. Use whey, milk, egg or milk and egg protein mix. All of these are highly digestible and will not cause intestinal problems. You can also use much less because of the purity of the product and how they are processed.

Method / Timing

Put the horse on six times the label recommendation for a human. Feed after work only on days the horse is exercised or raced.

SODIUM PHOSPHATE

Probability of Benefit	Dosage	Toxicity	Complementary Element	Usage
	20 grams daily		None	

Indications / Rationale

Sodium phosphate is used as a substitute for bicarbonate loading. Phosphate loading works extremely well in people but has equivocal results in horses. This may be at least partially due to the fact that many diets contain excessive amounts of calcium from either hay (alfalfa and other legumes) or water sources, potentially impairing uptake.

Phosphate loading has been reported in humans to buffer muscle acids (lower lactate), improve oxygen delivery to muscles and possibly enhance and improve use of glycogen.

Method / Timing

Give 3 to 4 grams daily, in divided doses, for three days prior to performance. The compound sodium phosphate must be fed at a rate of 20 grams per day to get that amount of phosphorous (the rest of the 20 gram weight is made up of the sodium component).

VITAMIN C

Probability of Benefit	Dosage	Toxicity	Complementary Element	Usage
	4.5 to 7.0 grams daily	Extremely high intake of vitamin C (probably 15 to 20 grams a day) can interfere with absorption of vitamin B12 from a horse's diet. However, a horse does not rely on diet to provide his B12 and may derive it from microorganisms in his intestine.	Vitamin E Bioflavinoids Manganese Copper Lipoic acid	

Indications / Rationale

Vitamin C is a water soluble (not stored in the body) antioxidant vitamin which is most intimately connected with the protection of the lungs and also functions as a necessary cofactor in the formation of healthy bones, joints, ligaments and tendons, as well as the formation of some hormones and other important substances. Vitamin C can also help reactivate vitamin E after it has neutralized free radicals. These electrically-charged molecules, created by exposure to chemicals, preservatives and pollution or as a waste product of exercise, attack normal body tissue to steal an electrical charge that re-stores them to normal. By doing so, they damage cells they attack, causing uncomfortable symptoms, such as inflammation, infections and, it's believed, permanent damage.

The generation of energy for work is associated with the production of large amounts of free radicals, especially when fat is being burned. Protects muscles from cellular damage from free radicals. Vitamin C/bioflavinoids protects lungs from irritants, infections and strengthens capillaries to help prevent lung bleeding.

Method / Timing

Vitamin C is available in a buffered form to help prevent stomach upset. Give 4.5 to 7.0 grams daily, double or triple this dose if there is a viral infection in the barn and for horses with soft tissue injuries or active arthritis.

VITAMIN E AND SELENIUM

Probability of Benefit	Dosage	Toxicity	Complementary Element	Usage
	Doses of E range from 500 I.U. to 5,000 for endurance racing and 3-day racers. One mg Super 1,000 IU of vitamin E.	The NRC has an upper safe limit of 10,000 IU/day of vitamin E for horses. The NRC recommends an upper safe limit of 2 ppm (2 mg for each kg of diet consumed), while many nutritionists recommend going as high as the 5 ppm mark.	Vitamin E and selenium complement each other and both complement vitamin C. Vitamin E complements zinc, while selenium complements copper.	

Indications / Rationale

Vitamin E helps with muscle weakness, cramping or tying-up, excessive inflammatory reaction to injuries or infections, impaired immunity. Selenium helps with muscular cramping, prevents the degeneration of some muscles, including the heart, and improves the oxidation of body fat stores.

All horses can benefit from vitamin E in their diet. Horses experiencing exercise-related muscle problems are prime candidates for selenium supplementation. Vitamin E levels in the diet are marginal at best, even for horses at maintenance, by NRC standards. Selenium is also borderline to deficient in most of the United States.

Method / Timing

For best absorption of selenium, use a supplement that contains selenomethionine (or selenocystine).

CHAPTER 6

RELIEVING HEALTH PROBLEMS THROUGH NUTRITION

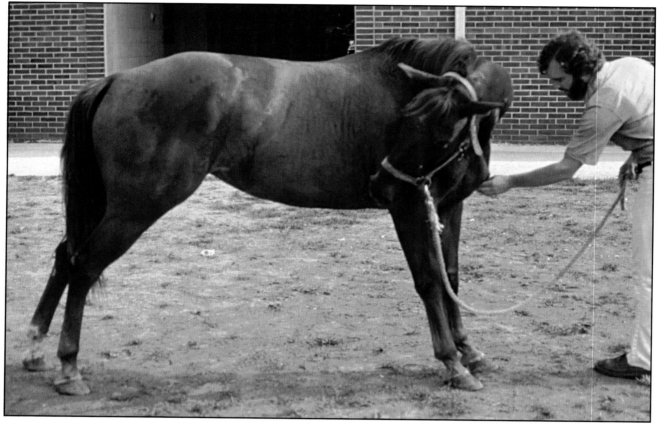

Many cases of colic respond to measures taken to encourage a healthy population of microorganisms in the intestine.

ABOUT THIS CHAPTER

This chapter will look at some common health problems in horses and how nutrition may play a role in the creation of such problems and/or in their treatment. It is NOT intended as a substitute for veterinary attention or concurrent treatment with any necessary drugs.

Although there are some disease states caused specifically by a nutritional inadequacy/deficiency, there are many more where poor nutrition is only part of the picture but will have an impact both on the likelihood of the problem developing and how

well/quickly the horse responds to medications and other treatment efforts.

For example, just about everyone is familiar with the latest rage in treatment of human colds—zinc lozenges. Zinc has important effects throughout both people's and horse's bodies in terms of strengthening the immune system and controlling inflammatory responses to invading organisms or tissue damage. Zinc can therefore make you, or your horse, better capable of fighting off a virus infection. At the same time, adequate zinc intake will help keep the severity of symptoms down. The same things can be said about vitamin C and viral infections. It is not at all unusual to hear that the incidence of common viral respiratory infections has been greatly reduced, if not eliminated, in barns where horses are being fed a properly supplemented diet.

While optimal nutrition will keep your horse as healthy as possible and speed response to treatment, it may not be enough to cure serious problems once they have become established. A horse with a broken bone will have the best possible healing when all major and minor minerals are being fed in correct balance. However, he will also need attention from an orthopedic surgeon.

A horse with OCD will have improved healing of the joints, better joint fluid and less pain on a supplement program of chondroitin sulfate/glucosamine, vitamin C and important trace minerals but will never be sound if there is a free floating piece of broken off cartilage in his joint. Returning to zinc, another common manifestation of inadequate zinc intake is poor hoof and skin quality. Supplementing zinc (and other needed nutrients) will resolve these problems but you will also have to treat any bacterial or fungal skin infections with the appropriate drugs and have your blacksmith deal with special needs for damaged feet in the interim.

This chapter will list some common health problems that are responsive to nutritional approaches, describe them as well as their symptoms, outline a possible supplement program, and give notes on treatment in general as well as on disease processes other than nutritional inadequacies that could be causing or contributing to the problem.

HEALTH PROBLEMS

The following health problems are grouped by type.
Here are some page numbers to help find specific problems:

Arthritis 176
Behavioral Problems (Nervousness,
    Aggression) and Problems with
    Mares in Season 178
Bloating/Excessive Gas 180
Bowed Tendon/Tendonitis 181
Brittle Feet/Cracking Feet 183
Chronic Colic/Indigestion 185
Constipation/Impaction 186
Desmitis/Suspensories/Curbs 187
Diarrhea 189
Dry Skin 191
Epiphysitis 192
Founder/Laminitis 193

Gingivitis (Gum Disease) 195
Heaves/COPD 196
Heel Scratches 197
Infertility 198
Lung Bleeding 200
Obesity/Weight Gain 202
Old Age 204
Osteochondrosis Dessicans–OCD 206
Pneumonia 208
Skin Infections 209
Tying-Up and Muscle Pain 211
Viral Respiratory Infections 213
Weight Loss/Failure to Gain Weight 214

ARTHRITIS

Symptoms

Joint swelling, lameness

Problem Description

Arthritis is an occupational hazard of active horses. Almost all horses show problems related to arthritis in their later years; far more quickly if heavily used.

Arthritis is an inflammation of the joint, which results in an imbalance between forces that break down/clean up cartilage and those that nourish/rebuild. Once started, the process may be difficult to stop by measures like rest alone.

Supplement Program

Nutrient	Dose	Comment
Glucosamine	9 grams/day	Controls inflammatory processes. Stimulates production of cartilage and hyaluronic acid. May be able to decrease dose to half or one third when horse stabilizes.
Chondroitin or Whole ground cartilage or perna mussel	7.5 grams/day 7-10 grams/day	Inhibits destructive enzymes. May enhance cartilage production.
Manganese	150 mg/day	Required for normal cartilage production. Co-factor for antioxidant enzymes.
Vitamin C	7-10 grams/day	Antioxidant for inflammatory control.
Bioflavinoids	22,000 mg/day	Antioxidants, enhance vitamin C effect.
Copper	75 mg/day	Essential for normal connective tissue formation.
Zinc	150 mg/day	Important for normal immune function. Co-factor for antioxidant enzymes.

Treatment Notes

Initial control may require use of either intravenous/intramuscular/intra-articular medications such as hyaluronic acid and low dose corticosteroids or PSGAGs (polysulfated glycosaminoglycans). However, the above supplement program assists greatly in freedom of movement, pain control and restoration of normal joint fluid.

Interval between symptom relapse to the point that injections are again required is greatly prolonged when using supplements (may go from every six weeks to as long as every six months, or longer, with continued use of the horse).

Pathology Discussion

Arthritis in horses is caused by a combination of factors, many of which are incompletely understood. What is clear is that mechanical stress on a joint can, over time, cause arthritis. The heavier the horse is worked, the earlier it may appear. Even slight conformation defects or shoeing imbalances greatly increase the risk for arthritis. When these latter two are present, the joint will not be loaded evenly. Changes seen when looking at such joints include thinning or holes in the cartilage, excessive bone deposition around the edges of overloaded areas and inflammation and thickening of the synovium, the lining tissue of the joint which secretes joint fluid.

BEHAVIORAL PROBLEMS (NERVOUSNESS, AGGRESSION) AND PROBLEMS WITH MARES IN SEASON

Symptoms

Overreacting to sound, jumpy when brushed or touched, rearing/biting/kicking/striking, resisting restraint or commands.

Problem Description

This category refers to horses that are difficult to get along with—nervous, aggressive, resistant to training, jumpy, hyperactive, etc.

Supplement Program

Nutrient	Dose	Comment
Thiamine	500 to 1,000 mg/day	B vitamin with calming properties in high doses.
Magnesium	2 to 10 grams/day	Major mineral, probably deficient in many diets either due to inadequate total magnesium in diet or diet too high in calcium.
B6	300 to 600 mg/day	B vitamin. Deficiency associated with irritability in other species.
Folic acid	6 to 12 mg/day	B vitamin. Adequate levels vital to production of normal levels of serotonin (a happy brain chemical).
Raw hemp or flax oil	3 to 6 tbsp/day	Essential fatty acids needed to maintain proper balance of hormones.

Treatment Notes

There is no substitute for proper training and sufficient exercise in controlling behavior problems of horses. Before blaming outside causes, have a professional evaluate your horse and how you interact with him/her. Feeding less grain does help make many horses calmer, for reasons that are not clearly understood. However, restricted feeding that actually causes a weight loss should be avoided. May substitute some fat calories for part of the grain calories but remember that pound for pound fat has 2.5 times as many calories as grain. High protein per se does NOT make horses hard to handle. However, most high protein diets are also high carbohydrate diets and this is the likely root of the problem.

Pathology Discussion

There are some horses that are just plain mean and resistant to training. However, most behavior problems have their roots in improper handling, training and management of the horse. On the nutritional front, have your diet checked for both total magnesium content and correct ratio of calcium:magnesium for good absorption (about 2 to 2.5:1).

Horses that are also thin and prone to colic or bouts of diarrhea could have B vitamin deficiencies. Feed a multivitamin supplement in addition to the higher levels of specific vitamins above.

Concerning problems with mares that are in estrus (heat), this is one of the most overtreated and unnecessarily treated problems in horses. The behavior of a mare in estrus is perfectly normal and should be accommodated as such. If you cannot deal with the behavior, don't have a mare! The above supplement program may help keep the mare on a more even keel during this time but will not change her normal cycling.

Continuation of a regular exercise program while avoiding the stress of shipping or competition is also helpful. However, the majority of mares continue to perform normally during this time, if they were well trained and schooled to begin with. Use of hormones can prevent estrus entirely but is expensive and may have long term health consequences.

Also, hormonal therapy to prevent estrus uses progesterone—the hormone that causes PMS in people. It is very unlikely that mares given progesterone actually feel better. Mares that become dangerously aggressive during estrus (a very small number) should be surgically spayed or sold to someone in a more appropriate situation.

BLOATING/EXCESSIVE GAS

Symptoms

Distention in the flanks, flatulence.

Problem Description

More of an annoyance than a serious problem in most cases but excessive gas production is a symptom of improper digestion. Flatulence accompanied by passage of small amounts of loose manure and fluid is common in horses that are excited (e.g., loading onto a van/trailer, at a race or show) and will resolve spontaneously.

Supplement Program

Nutrient	Dose	Comment
Diet modification, see treatment notes.		

Treatment Notes

Flatulence and bloating are usually diet related. Sudden change to alfalfa hay or a grain mix supplemented with alfalfa meal can cause this problem. Rapid addition of fat or supplements containing soybean (or grain mixes with soybean protein supplementation) may also be responsible. Yeast based supplements can even cause the problem in sensitive horses. Another prime offender is too much high quality pasture.

Pathology Discussion

Not associated with any disease state unless there are associated symptoms such as colic, diarrhea or weight loss.

BOWED TENDON/TENDONITIS

Symptoms

Lameness, heat, swelling, pain of tendon.

Problem Description

A true bowed tendon has one or more areas where the fibers of the tendon have actually been torn/ruptured. Tendonitis is an inflammation of a tendon, or its covering sheath, that may look exactly like a true bowed tendon and be extremely painful but does not actually involve a tear or rip in the tendon itself.

Supplement Program

Nutrient	Dose	Comment
Copper	75 mg	Important to formation of normal connective tissues. Most diets are deficient.
Zinc	150 mg	Co-factor to antioxidant enzyme systems. Most diets are deficient.
Manganese	150 mg	Co-factor to antioxidant enzyme systems.
Selenium	2 mg	Co-factor to antioxidant enzyme systems.
Raw hemp or flax oil	3-6 tbsp	Essential fatty acid source. Helps maintain normal balance between destructive and regenerative processes.
Vitamin C	7-10 grams/day	Antioxidant for inflammatory control.
Bioflavinoids	22,000 mg/day	Antioxidants, enhance vitamin C effect.
Methionine	2-4 grams/day	Sulfur containing amino acid essential to formation of strong tendons.
Lysine	2-4 grams/day	Essential amino acid.
Threonine	1-2 grams/day	Essential amino acid.
12% protein diet		Need high quality protein, such as alfalfa, soybean/milk mix or milk based protein.
B6	100-200 mg	Required for the proper use of protein in diet.
Biotin	2-5 mg	Important to health of all connective tissue structures.
Other B vitamins	See A-Z	All B vitamins complement each other's functions.

Treatment Notes

Use application of cold water/ice for first three days to control inflammation. Keep area bandaged for support and to control swelling. Do not buy into any miracle cures. Tendons that get better quickly were never truly bowed in the first place (tendonitis only was present). Do NOT feed supplemental iron. Iron is abundant in all equine diets and excess may interfere with absorption of other important minerals. Injection of the new drug BAP (beta-aminopropionitrill) helps prevent the formation of scarring between the tendon and sheath. Low level muscle electrical stimulation and electromagnetic therapy to injured tendons speeds healing and encourages regenerating fibers to align properly, resulting in a stronger repair. Most veterinarians also recommend low level daily exercise once the acute inflammatory stage is over (approximately two weeks)—usually hand walking—which also speeds healing and results in a stronger repair. Consult your veterinarian about other forms of ancillary therapy such as laser, ultrasound, magnetic therapy.

Pathology Discussion

Controlling the accumulation of fluid and blood inside an injured tendon is important to minimize scarring and speed healing. Tendons that are not injected with BAP or stimulated by light exercise, electrical stimulation, or electromagnetic therapy to their attached muscles during healing will have a higher percentage of fibers that are not properly aligned to accept stress.

BRITTLE FEET/CRACKING FEET

Symptoms

Dry, cracked and/or thin-walled feet which break up at the edges and hold shoes poorly.

Problem Description

Horses may also develop chronic foot soreness/tenderness.

Supplement Program

Nutrient	Dose	Comment
Copper	75 mg	Important to formation of normal connective tissues. Most diets are deficient.
Zinc	150 mg	Co-factor to antioxidant enzyme systems. Most diets are deficient. Known to be important to hoof health.
Manganese	150 mg	Co-factor to antioxidant enzyme systems.
Selenium	2 mg	Co-factor to antioxidant enzyme systems.
Raw hemp or flax oil	3-6 tbsp	Essential fatty acid source. Helps maintain normal balance between destructive and regenerative processes.
Vitamin C	7-10 grams/day	Antioxidant for inflammatory control.
Bioflavinoids	22,000 mg/day	Antioxidants, enhance vitamin C effect.
Methionine	2-4 grams/day	Sulfur containing amino acid essential to formation of strong hooves.
Lysine	2-4 grams/day	Essential amino acid.
Threonine	1-2 grams/day	Essential amino acid.
12% protein diet		Need high quality protein, such as alfalfa, soybean/milk mix or milk based protein.
B6	100-200 mg	Required for the proper use of protein in diet.
Biotin	2-5 mg	Important to health of all connective tissue structures.
Other B vitamins	See A-Z	All B vitamins complement each other's functions.

Treatment Notes

Nutritional inadequacies are often at the root of hoof problems. However, the problem is more complicated than just one or two nutrients. Although it will take 10 to 12 months to completely replace the hoof with a new, stronger wall, you should be able to see a difference in the quality of the newly grown hoof wall at the top within 2 to 3 weeks. Cracking and sore-footedness should also improve within a few weeks. It is also a good idea to have thyroid hormone levels checked on horses with bad feet. Underactive thyroids can cause this problem. If a thyroid problem is found, have your ration checked for iodine level. Do not supplement iodine without confirming the diet is low first. Iodine can be toxic. Talk to your vet or nutritionist.

Pathology Discussion

There are a variety of abnormalities found in feet, both on visual examination, under the microscope and using electron microscopes. No single vitamin or mineral accounts for all the abnormalities seen.

CHRONIC COLIC/INDIGESTION

Symptoms

Repeated problems with colic, usually low grade to moderate.

Problem Description

We are referring to horses with frequent bouts of mild to moderate abdominal pain which causes them to go off feed, paw, lie down, look at their flank and generally be uncomfortable but often with normal pulse and breathing rates, no other signs of a serious colic.

Supplement Program

Nutrient	Dose	Comment
Probiotics	Per directions	I have found that products containing bacterial growth enhancement factors are very effective with chronic colic. They encourage production of normal intestinal bacteria. If desired, you may begin the program by using live bacteria probiotic products. Continue use of products containing bacterial growth enhancement factors indefinitely.

Treatment Notes

Have a complete veterinary evaluation to rule out any serious or correctable cause for the abdominal pain, whether intestinal or not. Many cases respond to measures taken to encourage a healthy population of microorganisms in the intestine. Damage from intestinal parasites is another possible cause that may not be reversible. Keep the horse wormed regularly.

Pathology Discussion

With the exception of parasite damage, there is no specific intestinal lesion that can be found in most cases. A tumor or abscess in the abdomen can cause similar symptoms.

CONSTIPATION/IMPACTION

Symptoms

Decreased to no manure, straining, loss of appetite, may pass small amounts of liquid manure and/or small amounts of manure covered by a thick, rope-like mucus.

Problem Description

Caused by insufficient intake of water, excessively dry feeds, mechanical problems such as sand impaction, rubber fence impaction. May be the result of a complete obstruction which is a life threatening problem.

Supplement Program

Nutrient	Dose	Comment
Sodium chloride	1.5 ounce	Add to feed to encourage more water consumption.
Mineral oil	1 cup	Spray on hay. Lubricates the intestines.
Bran mash	3-4 pounds	Soak with equal volume of warm water until water is absorbed. Softens manure. Feed twice a week.
Psyllium	Per product instructions	Bulk laxative.

Treatment Notes

MUST have veterinary examination to rule out a more serious problem. Vet will treat with water and mineral oil by stomach tube to get things moving. Make sure horse is never out of clean water. Access to fresh grass helps very much. For persistent problems, consider switching to a complete pelleted feed and feeding this after soaking it in water.

Pathology Discussion

Damage from intestinal parasites may predispose to this problem by creating an area of bowel that does not move normally.

DESMITIS/SUSPENSORIES/CURBS

Symptoms

Lameness. Heat, swelling and pain of a ligament, such as the suspensory or a curb.

Problem Description

As with tendons, ligaments may either be only inflamed or inflamed and actually torn/ripped or ruptured.

Supplement Program

Nutrient	Dose	Comment
Copper	75 mg	Important to formation of normal connective tissues. Most diets are deficient.
Zinc	150 mg	Co-factor to antioxidant enzyme systems. Most diets are deficient.
Manganese	150 mg	Co-factor to antioxidant enzyme systems.
Selenium	2 mg	Co-factor to antioxidant enzyme systems.
Raw hemp or flax oil	3-6 tbsp	Essential fatty acid source. Helps maintain normal balance between destructive and regenerative processes.
Vitamin C	7-10 grams/day	Antioxidant for inflammatory control.
Bioflavinoids	22,000 mg/day	Antioxidants, enhance vitamin C effect.
Methionine	2-4 grams/day	Sulfur containing amino acid essential to formation of strong tendons.
Lysine	2-4 grams/day	Essential amino acid.
Threonine	1-2 grams/day	Essential amino acid.
12% protein diet		Need high quality protein, such as alfalfa, soybean/milk mix or milk based protein.
B6	100-200 mg	Required for the proper use of protein in diet.
Biotin	2-5 mg	Important to health of all connective tissue structures.
Other B vitamins	See A-Z	All B vitamins complement each other's functions.

Treatment Notes

As with tendons, rapid control of inflammation will speed healing and result in a more functional ligament after repair. Use cold water/ice to cool out the ligament.

Use bandaging to control fluid build up. Controlled exercise (hand walking) will assist healing.

Use of massage, warm hydrotherapy, laser, magnetic therapy, ultrasound and other ancillary therapies, as approved by your veterinarian, will assist in keeping blood flowing to ligaments.

Pathology Discussion

Ligaments have a poor blood supply, which limits both how quickly they can heal and how strong and functional the repair will be. Aggressive therapy (see treatment notes) and proper nutritional support are essential in preserving the athletic potential of the horse.

DIARRHEA

Symptoms

Abnormal amount and/or more fluid consistency to manure.

Problem Description

Diarrhea may be chronic (longstanding) or acute (sudden change). Secondary problems include loss of fluid (leading to dehydration), minerals and electrolytes.

Supplement Program

Nutrient	Dose	Comment
Probiotics	Per directions	I have found that products containing bacterial growth enhancement factors are very effective with chronic colic. They encourage production of normal intestinal bacteria. If desired, you may begin the program by using live bacteria probiotic products. Continue use of products containing bacterial growth enhancement factors until diarrhea has resolved (acute cases) or indefinitely.
Bran	1 pound	Judicious use of dry bran in small amounts may help restore a more normal character to the manure. Stop bran if no improvement in 24 hours.
B vitamins	See weight loss	Ration recommendations assume a certain level of production and absorption of B vitamins from the intestine. Diarrhea can be assumed to greatly decrease this.
Yogurt	1 cup	Nursing foals only. See Probiotics for older horses.
Vitamin C	7.5 grams	Use an esterified vitamin C product to avoid irritation.
Vitamin A	25,000 IU	Antioxidant.
Vitamin E	2000 IU	Antioxidant.
Potassium	4 grams	To replace losses.

Treatment Notes

Take the horse off grain, except for a tiny amount needed to give supplements or give supplements by oral syringe (mix in water or 1 to 2 ounces vegetable oil).

Diarrhea may be caused by sudden changes in feed (hay or grain portion), viruses, pathogenic bacteria or food sensitivities/allergies. Soy is a prime offender and is difficult to avoid unless you stick to plain grains. Calcium and magnesium supplements may be important in cases of chronic diarrhea. Feed at 6 to 8 grams of calcium and 3 to 4 grams of magnesium per day.

Pathology Discussion

Diarrhea caused by viruses, pathogenic bacteria or diseases of the bowel itself need veterinary attention. The dehydration that can accompany diarrhea may also require intravenous fluids or extra fluids by stomach tube.

Acute cases of diarrhea should be evaluated by a veterinarian. Chronic diarrhea associated with an obvious weight loss also requires veterinary attention.

DRY SKIN

Symptoms

Dry, flaking skin, poor hair quality.

Problem Description

Skin and coat lack the sleek, shiny appearance of health.

Supplement Program

Nutrient	Dose	Comment
Vitamin A	20,000 IU	May need up to 40,000 IU if using poor quality grass hay and no access to pasture.
Zinc	200 mg	Critical to health of the skin.
Raw hemp or flax oil	3-6 tbsp	Essential fatty acids (body cannot make them) that are required for skin health.
Biotin	2 mg	Required for health of all skin and connective tissues.
B6	200 mg	Needed for normal metabolism of protein. Supplement with B vitamins. All B vitamins work best when others are present in normal amounts.
Protein Minimum	12%	Need a high quality protein source such as alfalfa hay, alfalfa meal, soybean and milk protein mix or milk protein.
Lysine	2 grams	Essential amino acid.
Methionine	2 grams	Sulfur containing amino acid important to skin and connective tissue health.

Treatment Notes

Daily brushing and rubbing is essential to improve circulation and skin health. Bathe horse only if very dirty or sweaty. Use Ivory soap or Ivory flakes (dissolve first in hot water) and rinse very thoroughly.

Pathology Discussion

Skin reflects the nutritional status of the entire body. Dry skin is not a disease per se but a symptom of nutritional inadequacy.

EPIPHYSITIS

Symptoms

Swelling, heat and pain at actively growing joint.

Problem Description

Epiphysitis is an inflammation of the growth plate— the region where bone grows.

Supplement Program

Nutrient	Dose	Comment
Balanced Diet		Diet must be balanced for both the age and the estimated adult weight of the horse. Diet must have adequate but not excessive energy. Protein needs must be met by high quality protein. Ensuring correct mineral intake is critical.

Treatment Notes

Exercise is restricted during an epiphysitis problem. The animal should be kept alone in either a very roomy stall or a small paddock/pen to avoid aggravating the area by excessive playing/running. Get an expert nutritional consultation for diet evaluation and follow the recommendations to the letter.

Pathology Discussion

Epiphysitis is *NOT* caused by too high a protein content in the diet. Adequate high quality protein is a must for proper bone growth and formation. Do not feed a low protein diet. High calories, rapid growth, rapid weight gain are all contributing factors. Conformation faults may also put uneven stresses on growth plates, causing inflammation. Excessive rough housing may cause injury and result in epiphysitis.

FOUNDER/LAMINITIS

Symptoms

Reluctance or refusal to walk. Feet hot to the touch. Extreme pain.

Problem Description

Laminitis is an inflammation of the live tissues inside the hoof. These tissues form a connection between the hoof wall and the coffin bone inside. The build up of fluid and blood that occurs inside the foot with laminitis is extremely painful—equivalent to the pain experienced if you close a car door on your fingertip/nail but much worse since the horse must also put his weight on the injured area.

Supplement Program

Nutrient	Dose	Comment
Copper	75 mg	Important to formation of normal connective tissues. Most diets are deficient.
Zinc	150 mg	Co-factor to antioxidant enzyme systems. Most diets are deficient. Known to be important to hoof health.
Manganese	150 mg	Co-factor to antioxidant enzyme systems.
Selenium	2 mg	Co-factor to antioxidant enzyme systems.
Raw hemp or flax oil	3-6 tbsp	Essential fatty acid source. Helps maintain normal balance between destructive and regenerative processes.
Vitamin C	7-10 grams/day	Antioxidant for inflammatory control.
Bioflavinoids	22,000 mg/day	Antioxidants, enhance vitamin C effect.
Methionine	2-4 grams/day	Sulfur containing amino acid essential to formation of strong hooves.
Lysine	2-4 grams/day	Essential amino acid.
Threonine	1-2 grams/day	Essential amino acid.
12% protein diet		Need high quality protein, such as alfalfa, soybean/milk mix or milk based protein.
B6	100-200 mg	Required for the proper use of protein in diet.
Biotin	2-5 mg	Important to health of all connective tissue structures.
Other B vitamins	See A-Z	All B vitamins complement each other's functions.

Treatment Notes

Veterinarian and blacksmith are equally important to proper treatment.

Involve both simultaneously. Aggressive therapy will enhance chances of a happy outcome. Seek out and use experts in dealing with this problem. Treat acute cases involving hot feet with cold water/ice soaks. The old time remedy of standing the horse in a running stream is still a good idea. Mild exercise (which must be determined on an individual basis depending on what X-rays show—see pathology discussion) may help in later stages of healing.

Pathology Discussion

The coffin bone may be pulled loose from some or all of its attachments and rotate inside the hoof wall. In extreme cases, it can put enough pressure on the sole of the foot that the bone actually breaks through. The horse may also lose his hoof wall.

GINGIVITIS (GUM DISEASE)

Symptoms

Reddened, possibly bleeding gums. Foul odor on breath. Gums recede away from the roots of the teeth.

Problem Description

Gingivitis, inflammation/bacterial infection of the gums, is a common problem in older horses but can affect horses of any age. Eating is probably painful, leading to incomplete chewing of feeds and predisposing the horse to inadequate digestion and problems such as choke (food caught in the esophagus).

Supplement Program

Nutrient	Dose	Comment
Coenzyme Q_{10}	250 mg	Add to small amount of vegetable oil (1 to 2 ounces) and mix into feed immediately before feeding or give by an oral syringe. Degrades quickly, especially on exposure to light.
Vitamin C	7.0 grams	Antioxidant, anti-inflammatory and supports immune system. Builds strong connective tissues.
Vitamin A	25,000 IU	Antioxidant.
Glucosamine	9.0 grams for two weeks then 3.0 grams	Important to connective tissue health.
Zinc	250 mg	Helps support immune system and antioxidant systems.
Copper	75 mg	To balance zinc.

Treatment Notes

The goal is to attack inflammation and infection in the gums. Levels of coenzyme Q_{10} decrease with age. This nutrient is extremely helpful in combating gum disease.

Other nutrients serve to improve antioxidant function and boost the immune system.

Pathology Discussion

Health of the gums reflects overall nutrition as well as vitamin status. The ability of the mucus membranes, such as the gums, to resist inflammation and infection is strongly affected by antioxidant status.

HEAVES/COPD

Symptoms

Decreased exercise tolerance, cough, elevated respiratory rate, difficulty catching breath.

Problem Description

Heaves (chronic obstructive pulmonary disease) has many parallels with asthma, allergic bronchitis and, in the final stages, emphysema in people. The name heaves comes from observing how hard the horse is working to breathe (sides heaving) during an attack.

Supplement Program

Nutrient	Dose	Comment
Vitamin C	4.5 grams	Boosts immune system, antioxidant.
Bioflavinoids Complex with Hesperidin	22,000 mg	Antioxidant, complement vitamin C.
Vitamin E	2,000 IU	Antioxidant, aids healing.
Selenium	2 mg	Complements vitamin E, antioxidant.
CoEnzyme Q_{10}	200 mg	Give twice a day for 10 days, then once a day. Antioxidant, improves oxygen use of tissues and eases work of breathing.
Zinc	100 mg	Three times a day until symptoms gone, then 150 mg/day.
Grapeseed extract	0.75 ounce	Twice a day until symptoms gone, then daily.

Treatment Notes

Lung infections and exposure to materials in a stable environment (dust, hay and straw dust, microscopic molds, irritating gases such as ammonia from urine breakdown, etc.), as well as pollutants in the air and allergies to plant and tree pollens may all play a role in causing this problem in sensitive horses.

Most horses do better if kept outside. When inside, good ventilation is absolutely essential, even in winter. Wet hay before feeding. Use only dustfree grains or moisten with oil or molasses. Bed on wood chips. Prescription medications will be needed for moderate to severe cases, at least during "attacks."

Pathology Discussion

This disease eventually produces irreversible changes in the lungs such as thickening of the small airways (bronchi and bronchioli), collapse of airways, rupture of areas where gas exchange (oxygen in, carbon dioxide out) occurs.

HEEL SCRATCHES

Symptoms

Cracked, open skin on heels and back of pastern. Severe cases may bleed.

Problem Description

Often begins as a mechanical irritation from plants or dirt. High moisture in this area worsens the problem. Secondary infections are common.

Supplement Program

Nutrient	Dose	Comment
Vitamin A	20,000 IU	May need up to 40,000 IU if using poor quality grass hay and no access to pasture.
Zinc	200 mg	Critical to health of the skin.
Raw flax or hemp oil	3-6 tbsp	Essential fatty acids (body cannot make them) that are required for skin health.
Biotin	2 mg	Required for health of all skin and connective tissues.
B6	200 mg	Needed for normal metabolism of protein supplement with B vitamins. All B vitamins work best when others are present in normal amounts.
Protein	Minimum 12%	Need a high quality protein source—alfalfa hay, alfalfa meal, soybean and milk protein mix or milk protein.
Lysine	2 grams	Essential amino acid.
Methionine	2 grams	Sulfur containing amino acid important to skin and connective tissue health.

Treatment Notes

Local care important to clearing up problems. Supplements help keep it from coming back. Mix up a cream using 12 oz of petroleum jelly (or whipped Vaseline), 1,000 units vitamin E, 400 units vitamin A, 8 cc of vinegar, 8 cc of 20% sodium iodide solution (from your vet), 3 grams of glucosamine. May add 10 mg of prednisone or prednisolone if area is extremely inflamed.

Gently cleanse skin using an iodine based soap (small amount) on the palm of your hand (no sponge or brush). Apply cream to area at night and after work or bathing. Keep bandaged. Complete healing in as little as two to three days.

Pathology Discussion

Skin that is weakened or lacks sufficient normal immune responses is prone to this type of problem.

INFERTILITY

Symptoms

Low sperm counts or inability to conceive or hold a pregnancy, low libido, weak estrus periods.

Problem Description

Infertility in stallions may be related to either lack of desire to perform, defective sperm, low sperm counts or low sperm motility. Some stallions have sperm that does not survive freezing processes required for artificial insemination. In mares, the problem may be either an inability to conceive or inability to hold onto the pregnancy.

Supplement Program

Nutrient	Dose	Comment
Gamma oryzanol oil emulsion*	1,000 mg	Natural plant hormone with proven anabolic (body building) effects, which is also gaining recognition as being helpful in infertility in stallions, related to low sperm counts and in helping mares to both conceive and carry to full term.
Iodine	1 to 2 mg/day	Important in maintaining normal thyroid function and fertility.
Selenium	1 to 2 mg/day	Antioxidant. Deficiency associated with infertility in other species.
Beta-Carotene & vitamin A	250 mg/day 10,000 IU/day	Supplementation associated with improved mare fertility in early part of breeding season.
Zinc	150 mg/day	Linked to male impotence and infertility in other species.
Copper	50 mg/day	Deficiency linked to weakness of uterine arteries.
Vitamin C	4.5 grams/day	Improves sperm counts and viability in other species.
Multivitamin and mineral supplement		Support general health with an all purpose supplement matched to your diet.

* Do not use gamma oryzanol powders or gamma oryzanol in oil base that is not an emulsion. There are many copy cat products which actually contain very very little gamma oryzanol. Ask manufacturer to supply you with laboratory analysis proving content of gamma oryzanol is as stated.

Treatment Notes

Many cases of infertility are caused by stress. Maintaining animals in optimal nutritional condition (sleek but not fat, coat shining) enhances fertility as well as general health. Avoid all mental stress. Animals previously given synthetic anabolic steroids (e.g., Equipoise, Winstrol) may be infertile for a year or more. Anatomical abnormalities and other physical causes, such as infection, must be ruled out.

Pathology Discussion

Most cases of infertility are caused by stress, hormonal imbalances and low grade infections. Abnormalities of the testicles, ovaries, uterus or other portions of the reproductive tract are rarely seen.

LUNG BLEEDING

Symptoms

Decreased exercise tolerance, coughing after work, blood seen at nostrils or during examination of lungs with an endoscope.

Problem Description

First symptom is usually decreased ability to perform hard/fast work. Bleeding from lungs occurs at high speeds/high heart rates.

Supplement Program

Nutrient	Dose	Comment
Vitamin C	4.5 grams	Boosts immune system, antioxidant
Bioflavinoids Complex with Hesperidin	22,000 mg	Antioxidant, complement vitamin C.
Vitamin E	2000 IU	Antioxidant, aids healing.
Selenium	2 mg	Complements vitamin E, antioxidant.
CoEnzyme Q_{10}	200 to 400 mg	Antioxidant, improves oxygen use of tissues and eases work of breathing. Give twice a day for 10 days then once a day.
Zinc	100 mg	Three times a day until symptoms gone then 150 mg/day.
Grapeseed extract	0.75 ounce	Twice a day until symptoms gone then daily.

Treatment Notes

The drug Lasix (furosemide) helps most horses with lung bleeding. It is a diuretic but that is not the way it helps. Lasix actually decreases pressures on the right side of the heart (the side that pumps through the lungs).

There are many potential side effects to use of Lasix, related to the loss of electrolytes in the urine and loss of too much fluid from the body. To prevent these, follow these rules:
• Do NOT restrict water intake on day of race.
• Give 4 to 5 grams of potassium salt and 4 to 5 grams of magnesium salt the night before the race. If horse is de-hydrated by skin pinch test, do not give these. Have horse tubed with several liters of water in addition to the K and mg salts.

Using the above supplement program, many horses have improved beyond the improvement they received from Lasix alone, and some can be taken off Lasix entirely.

Feeding vitamin K will NOT help a bleeder. Injectable vitamin K is highly toxic (and won't help either).

Pathology Discussion

Lung bleeding is caused by very high pressures in the tiny blood vessels inside the lungs. It can happen during extreme exercise even in a normal horse but allergies and prior lung infections may make a horse more likely to have lung bleeding.

OBESITY/WEIGHT GAIN

Symptoms

Persistent abnormally high weight despite feeding in appropriate amounts.

Problem Description

Some horses, because of their basic metabolism or lack of sufficient exercise, will gain excessive amounts of weight as fat. This makes normal breathing difficult and also places tremendous strains on the spine and joints.

Supplement Program

Nutrient	Dose	Comment
Minerals	Calcium 8 grams Phosphorus 5 grams Potassium 1.5 grams Copper 150 mg Zinc 350 mg Magnesium 4 grams Manganese 275 mg Selenium 2.5 mg	Additional minerals are always needed on reduced calorie diets. The list shown here gives dosages for horses on a grass hay diet. Needs will vary with different hay types and with different qualities of hay.
Vitamin A	40,000 IU	Will not need with alfalfa hay.
Vitamin E	2,000 IU	
Lysine	4.0 grams	
Methionine	2.0 grams	

B vitamins	Minimums:
	  Riboflavin 50 mg
	  Niacin 200 mg
	  Choline 250 mg
	  D-pantothenate 90 mg
	  Pyridoxine 50 mg
	  Thiamine 200 mg
	  Folic acid 30 mg
	  Biotin 5 mg	

Vitamin C	4.5 grams	
Fatty acids	4 to 6 tablespoons of unprocessed vegetable oil	This is provided to meet minimal fat needs. Do not use vegetable oils purchased in a grocery store. The processing destroys the essential fatty acids.

Treatment Notes

Weight reduction requires a combination of caloric restriction and regular exercise. Drastic cuts in feed will not get the job done and can be life threatening in ponies. Eliminate grain. Use carrots for snacks. Weigh your hay carefully. Feed 0.75% of horse's weight in hay per day. Use a clean, late cutting grass hay. Alfalfa has more calories than grass hay and you will not be able to feed as much. Bed the horse on sawdust or wood shavings.

A hungry horse will fill up on straw which has very low vitamin and mineral levels but almost as many calories as hay. Turn out in a dirt paddock and/or limit grazing to 1 hour per day. Daily exercise, even on a lunge line, is a must. Aim for at least 20 minutes at the trot, starting at 10 minutes if the horse cannot tolerate this initially. Exercise is just as important as calorie restriction.

Pathology Discussion

On rare occasions, excessive weight gain may be related to low thyroid hormone levels. Do not begin thyroid supplementation without having low levels of the hormones confirmed by blood tests. If needed, use only the amount recommended by your veterinarian. Periodic checks (at least every 3 to 4 weeks) on thyroid hormones must be done while the horse is being supplemented.

OLD AGE

Symptoms

Anyone can recognize an old horse but what constitutes old will vary from animal to animal. An improperly managed race horse may look old when he is 5, while ponies and cross bred horses are often active, vital and young looking into their 20s or 30s.

Problem Description

Aging is not a disease, it is a natural and inevitable process. However, aging is one of the hottest areas of research at this time and scientists are working hard to determine what makes the body age and how to postpone it. There are biological clocks, probably genetically programmed, that signal how fast to grow, when to stop growing, when to become sexually mature (as well as when to stop functioning sexually), etc. When major hormone systems shut down, other hormone systems are also affected, as is the metabolism of the entire body. Aging also occurs as damage to vital organs (skin, heart, lungs, liver, kidneys, etc.) accumulates over time. The external effects of toxins, drugs and radiation damage the body, but even exercise takes its toll—all acting by the same mechanism, the generation of free radicals. The combination of digestive processes that are not as efficient as they were in young animals with inadequate nutrient levels in the diet, decreased ability to chew food and a decreased appetite also contribute to aging and weakening of the tissues.

Supplement Program

Nutrient	Dose	Comment
Linoleic & linolenic acid*	3 to 6 tbsp/day	Helps balance inflammatory reactions, health of skin and feet, hormone production.
CoQ$_{10}$	180 to 500 mg/day	Potent antioxidant, improves ability of the cells to use oxygen and generate energy.
Multivitamin and Mineral with: Pyridoxine Folic acid Vit E Vit C Copper Cobalt Zinc Manganese Lysine Pantothenate Riboflavin Vit A Vit D3 Niacin Selenium Methionine Threonine Calcium	 25 mg/day 25 mg/day 500+ IU/day 4.5 gm/day 60 mg/day 20 mg/day 180 mg/oz 100 mg/oz 2 gm/day 65 mg/day 40 mg/day 20,000 IU/day 5,000 IU/oz 125 mg/oz 2 mg/day 2 gm/day 1 gm/day 6 gm/day	Antioxidant protection, supplementation to combat probable decreased absorption and/or inadequate intake. Research has shown that older horses have less efficient absorption of phosphorus. Supplements containing phosphorus are indicated, and using one with a yeast base may also help as yeasts have been shown to improve mineral absorption. Blood vitamin C levels have also been demonstrated to be low in older horses—solid reason for supplementing with C and the other antioxidants. You should also choose a supplement that contains chelated minerals—minerals bound to an organic protein or polysaccharide—instead of inorganic mineral salts such as sulfates or oxides (e.g., copper proteinate instead of copper sulfate). Minerals in this form are much more efficiently absorbed.

Supplement Program – continued

Nutrient	Dose	Comment
Phosphorus Magnesium Gamma Oryzanol	3 gm/day 3 gm/day 1,000 mg/day	Natural anabolic without anabolic side effects.

* e.g., unprocessed ("raw") flaxseed oil—see Essential Fatty Acids in Chapter 4, A to Z Other for sources.

Treatment Notes

The above recommendations are a sample program for maximizing protection from free radicals, combating decreased digestive efficiency and providing a little insurance against dietary insufficiencies. No single product is going to contain all these nutrients in these specific proportions. However, your horse might not need them all—or might need more. The older horse also needs an easily digestible diet with very high quality protein. Ingredients such as beet pulp and brewer's byproducts are easily digested and are more concentrated sources of calories. Grains such as corn and oats should be processed (rolled, steam crimped, etc.). Extruded feeds are often advertised as good for horses, but you have to feed a larger volume, which often does not work well with the limited appetite of an older horse. Mashes/gruels will be needed for horses with missing or few teeth. There are several specific senior diets on the market that address the older horse's special needs. For specific recommendations on supplementing the diet of your older horse, consult with a veterinarian or nutritionist who specializes in this problem.

Pathology Discussion

We covered the basics of how aging occurs above. To keep your horse young, feed a very high quality diet at all times, have your diet analyzed by a specialist to ensure it is adequate (if not optimal) and supplement with antioxidant nutrients such as vitamin E, C and A, selenium, copper, manganese and zinc, as needed.

OSTEOCHONDROSIS DESSICANS – OCD

Symptoms

Lameness and swelling in one or more joints.

Problem Description

OCD is a congenital disease that results in improper formation of the joint cartilage. Excessive joint swelling may be present from an early age. Any joint may be involved. Lameness usually does not develop until the animal is put to work. There is very strong evidence to suggest a nutritional component (pregnant mare and young foal) to OCD. Hereditary factors may also be at work.

Supplement Program

Nutrient	Dose	Comment
(Mature, nonpregnant horses)		
Copper	75 mg	Low copper implicated in OCD. Most diets are deficient.
Zinc	150 mg	Low zinc implicated in OCD. Most diets are deficient.
Manganese	150 mg	Essential to normal cartilage formation.
Raw flax or hemp oil	3-6 tbsp	Essential fatty acid source. Helps maintain normal balance between destructive and regenerative processes.
Vitamin C	4.5-7.0 grams/day	Antioxidant for inflammatory control.
Bioflavinoids	22,000 mg/day	Antioxidants, enhance vitamin C effect.
Glucosamine	9 grams/day	Controls inflammatory processes. Stimulates production of cartilage and hyaluronic acid. May be able to decrease dose to half or one third when horse stabilizes.
Chondroitin sulfate or Whole ground cartilage or Perna mussel	7.5 grams/day 7-10 grams/day	Chondroitin 7.5 grams/day inhibits destructive enzymes. May enhance cartilage production.

Treatment Notes

May require injectable therapy as described for arthritis to affect good control. However, there is little to no role for corticosteroids in the treatment of this disease. Current surgical approaches smooth off or remove loose pieces of cartilage but there is no solid evidence they produce much better results than medical management. Keep alert for developments in surgery that involve placement of cartilage grafts into involved joints. This is the most exciting and promising prospect for the future.

Pathology Discussion

OCD results in such problems as thin or weak cartilage, cartilage flaps or free floating pieces, bone cysts underlying the cartilage.

PNEUMONIA

Symptoms

Fever, moist cough, difficulty breathing, loss of appetite, low exercise tolerance, white to yellow nasal discharge

Problem Description

Pneumonia is infection of the lungs. Viral illnesses may cause a relatively mild and short lived pneumonia but set the stage for the more serious bacterial or fungal pneumonias to develop.

Supplement Program

Nutrient	Dose	Comment
Vitamin C	4.5-7.0 grams	Boosts immune system, antioxidant.
Bioflavinoids Complex with Hesperidin	22,000 mg	Antioxidant, complement vitamin C.
Vitamin E	2,000 IU	Antioxidant, aids healing.
Selenium	2 mg	Complements vitamin E, antioxidant.
CoEnzyme Q_{10}	400 mg	Antioxidant, improves oxygen use of tissues and eases work of breathing.
Zinc	100 mg	Three times a day until symptoms gone then 150 mg/day.
Grapeseed extract	0.75 ounce	Twice a day until symptoms gone then daily.

Treatment Notes

Bacteria pneumonias MUST also be treated with appropriate antibiotic and/or antifungal drugs under a veterinarian's direction.

Pathology Discussion

Pneumonia is a very serious condition and can permanently decrease the horse's exercise capacity. Pneumonia may scar the lungs and may predispose the horse to later problems such as emphysema/heaves or lung bleeding during exercise.

SKIN INFECTIONS

Symptoms

Bumps and crusts on skin. May have areas of hair loss. Areas under crusts/scabs is moist and oozing. Fungal infections typically have very thick crusts.

Problem Description

Skin infections are most common on the legs but can be found anywhere on the body. In many cases they are indicative of a weak immune system and poor nutritional state.

Supplement Program

Nutrient	Dose	Comment
Vitamin A	20,000 IU	May need up to 40,000 IU if using poor quality grass hay and no access to pasture.
Zinc (with 65 mg copper to balance)	200 mg	Critical to health of the skin.
Raw hemp or flaxseed oil	2 oz	Essential fatty acids (body cannot make them) that are required for skin health.
Biotin	2 mg	Required for health of all skin and connective tissues.
B6	200 mg	Needed for normal metabolism of protein. Supplement with B vitamins. All B vitamins work best when others are present in normal amounts.
Vitamin C	4.5 grams	Boosts immune system, antioxidant.
Bioflavinoids	22,000 mg	Antioxidant, complement vitamin C.
Vitamin E	2,000 IU	Antioxidant, aids healing.
Selenium	2 mg	Complements vitamin E, antioxidant.
Protein	Minimum 12%	Need a high quality protein source such as alfalfa hay, alfalfa meal, soybean and milk protein mix or milk protein.
Lysine	2 grams	Essential amino acid.
Methionine	2 grams	Sulfur containing amino acid important to skin and connective tissue health.
MSM	10 gm	Organic sulfur source for connective tissue health.

Treatment Notes

Local care important to clearing up problems. Supplements help keep it from coming back. Mix up a cream using 12 oz of petroleum jelly (or whipped Vaseline), 1,000 units vitamin E, 400 units vitamin A, 8 cc of vinegar, 8 cc of 20% sodium iodide solution (from your vet), 3 grams of glucosamine. May add 10 mg of prednisone or prednisolone if area is extremely inflamed.

Gently cleanse skin using an iodine-based soap (small amount) on the palm of your hand (no sponge or brush). Apply cream to area at night and after work or bathing. Keep bandaged using a sweat wrap (cream, cotton wrap, plastic wrap, outside wrap). Complete healing in as little as 2 to 3 days. Infections often arise inside sponges and bathing equipment. Disinfect buckets with chlorine bleach. Throw out sponges and buy new ones. Wash involved areas using your hands only.

Pathology Discussion

Skin that is weakened or lacks sufficient normal immune responses is prone to this type of problem. Long winter coats may predispose to skin infections.

TYING-UP AND MUSCLE PAIN

Symptoms

Muscle spasm, pain on palpation of muscles, reluctance to move, may have dark brown colored urine.

Problem Description

Muscle pain from overuse or other causes may occur anywhere on the body. Tying-up refers to a syndrome of extreme muscle spasm and pain that is triggered by exercise. The brown urine is caused by pigments from damaged muscle leaking out into the blood and being cleared by the kidneys.

Supplement Program

Nutrient	Dose	Comment
Magnesium	4.0 grams/day	May need more if feeding alfalfa hay. Mineral critical to normal function of muscle cells.
Calcium	2.0 grams/day	Do not use with alfalfa hay.
Potassium	4.0 grams/day	
Zinc	200 mg	Important to normal metabolism of carbohydrates and cofactor for antioxidant enzyme systems. Give 70 mg copper to balance.
Vitamin E	2,000 to 5,000 IU	Important intracellular antioxidant—controls damage to muscle cells.
Vitamin C	4.5 - 7 grams	Recycles used vitamin E.
Selenium*	4 mg	Another vital antioxidant, complements the vitamin E.
Branched chain amino acids	25 to 30 grams	Mixture of leucine, isoleucine and valine. Available as a paste (PRO-BURST™). Give 30 minutes before exercise, after exercise and if horse has an attack. Critical amino acids for working muscles.
Carnitine	6 grams	Probably must give daily to get any benefit. Important to metabolism of fats and proteins.
Lipoic acid	1,000 mg	Potent antioxidant. Also improves metabolism inside muscle cells. Is used day before and day of important competitions. Best effect if used daily.

*Get veterinarian to evaluate Selenium level in diet before using this dose.

Supplement Program – continued

Nutrient	Dose	Comment
DMG (Dimethylglycine)	1,500 mg	Feed twice a day for 10 days then once a day. Improves metabolism of muscle cells but only in horses that are not yet fit for their work load.

B Vitamins
    B1
    B2
    B3
    B6
 	
Pantothenate
Choline
Inositol
Folic acid
	
	2,000 mg
	1,000 mg
	3,000 mg
	1,000 mg
	
	2,000 mg
	2,000 mg
	1,000 mg
	6 mg
		Give this amount for first three days of treatment or first three days after an attack then decrease to ° these doses to ˇ these doses (smaller horse) for daily supplementation. Adequate B vitamin levels are necesssary for correct use of fuels by muscle cells.

CoEnzyme Q_{10}	400 mg	Antioxidant and improves utilization of oxygen by muscle cells. Must use for at least 3 days at double this dose before a competition or use daily at this dose.

Treatment Notes

The above program is very helpful in eliminating chronic muscle soreness. Horses with persistent pain, especially along backs, may have an underlying spinal problem. Even these will respond if treatment with the prescription product RVI is also used.

Tying-up has many possible causes, some of which are basic defects in the metabolism of the horse. This program is very effective for 95% of horses. Always also check thyroid function. Low or high thyroid hormone levels can cause tying-up. Of the 5% that do not respond, an inherent defect in energy generation in the muscle is likely.

Many horses respond to elimination of grain from their diet. However, unless the horse has been biopsied and proven to have a storage disease that indicates he has a genetic defect causing problems with carbohydrates, eliminating grain is not necessary when the above supplements are used. Eliminating grain will adversely affect racing performance.

Pathology Discussion

Muscle spasm, especially tying-up, is an energy crisis. The muscle cell does not have enough energy to relax and is trapped in a contracted state.

VIRAL RESPIRATORY INFECTIONS

Symptoms

Cough, fever, clear nasal discharge.

Problem Description

Viral respiratory infections are the equivalent to our colds and flu.

Supplement Program

Nutrient	Dose	Comment
Vitamin C	4.5 grams	Boosts immune system, antioxidant, give three times a day until symptoms gone then once a day.
Bioflavinoids	22,000 mg	Antioxidant, complement vitamin C.
Grapeseed extract	0.75 ounce	Potent anti-inflammatory.
Vitamin E	2,000 IU	Antioxidant, aids healing.
Selenium	2 mg	Complements vitamin E, antioxidant.
Vitamin A	25,000 IU	Needed for healing and health of lining of respiratory tract.
Zinc	100 mg	Three times a day until symptoms gone then 150 mg/day.
L-lysine	4 gm	Helps in some viral infections.

Treatment Notes

Prevention using appropriate supplements works much better than treatment after the fact. Horses supplemented as above have far, far fewer respiratory infections and the few they may have are generally very mild.

Pathology Discussion

Viral infections weaken the local immune responses, setting the stage for bacterial infections and/or pneumonia. Viral infections can also involve the muscles, including the heart.

WEIGHT LOSS/FAILURE TO GAIN WEIGHT

Symptoms

Normal to increased appetite but failure to hold weight well.

Problem Description

Horses with difficulty gaining and holding weight are often highly strung pure-breds. Difficulty with weight gain may also be the result of insufficient feed in relation to the amount of work being performed or poor quality feed.

Supplement Program

Nutrient	Dose	Comment
Fat	10% of total ration	Vegetable oils are more natural for the horse than animal fat products although horses do digest and use processed animal fats well. Long term health consequences of feeding animal fats are unknown. All vegetable oils are palatable but best is a blend of coconut and soy oil.
Probiotics	Per directions	Probiotics are substances that either contain live, beneficial bacteria and/or substances that encourage their growth in the intestinal tract. These are often added to other supplements or may be purchased separately. Do NOT use iron containing products in horses under six months of age.
Yeast	Per directions	Yeast has a beneficial effect on digestion in the large intestine and improves availability of some minerals. It is of potential benefit in horses with poor digestion but not a primary treatment for poor weight gain.
Multivitamin	Per directions	Vitamin supplementation is especially important in thin horses. Choose one with generous amounts of B vitamins.
Thiamine	250 to 1,000 mg/day	Especially valuable with thin horses that are nervous.

Treatment Notes

Begin with high quality diet in adequate amounts. Horses at maintenance require about 1% of their body weight in feed on a daily basis, more for nervous horses or hard keepers. Addition of fat helps boost calories without adding bulk. Horses in hard work will need about 2% of their body weight in feed daily, half or more as concentrate (grain) in many cases. Institute and keep a good worming program, using a larvicidal wormer (ivermectins, Quest or triple dose fenbendazole) initially, followed by worming at 3 to 4 week intervals. Any of the common wormers may be used but best results will be obtained with ivermectin products or Quest. Quest should not be used regularly in horses less than a year old as it is not proven effective against some parasites of very young horses. Introduce fats gradually, starting at 2 oz per day and increasing by 1 to 2 oz every 2 to 3 days.

Pathology Discussion

If weight loss has not been preceded by an increase in work level or a change to lesser quality feed, have the horse checked out for possible digestion problems. This is especially important if the horse has not had this problem in the past. Hyperthyroidism, usually caused by overuse of thyroid supplements, will also cause weight loss. Intestinal parasites interfere with digestion and absorption in many ways.

CHAPTER 7

CONSUMER'S GUIDE TO SUPPLEMENTS

Feeds and supplements are available in a variety of textures.

ABOUT THIS CHAPTER

To use this book to your greatest advantage you will need a plan. The following chapter is a step-by-step guide on how to use the information in the other chapters as a guide to determining which nutrients you need to supplement and how to find a supplement that meets those needs.

DETERMINING YOUR SUPPLEMENT NEEDS

GETTING A ROUGH IDEA

To begin determining your supplement needs, go to Chapter 1, Basic Nutrition, and locate your horse by level of use (maintenance, light work, etc.). Locate your basic type of diet by hay type and whether or not you feed grain. If your horse is receiving a complete feed, check the label list of ingredients to determine the type of hay or other roughage (usually beet pulp) in the product. Then, look under the hay and grain diet that matches the hay type in the complete feed.

Using the diet charts, locate nutrients coded as red (deficient) or yellow (borderline). These will form the core of your required nutritional supplements.

Please remember that the numbers used to create these charts are only the averages as published by the NRC and may vary quite widely from the actual analysis of hay and grain from your region. (See Factors Affecting Mineral Levels in Feeds, Chapter 1, Basic Nutrition.) The basic diet charts also used 50:50 grain and hay diets. If you feed less grain, your numbers will be different. To see how the level of grain decreases or increases the level of a certain nutrient, compare the mixed grain and hay diet with the hay only charts. If a nutrient changes from red or yellow to green in the hay only diet, that nutrient is higher in the hay than in the grain and vice versa. More information about the specific nutrient levels in your local hays and grains can be obtained from a feed analysis (see below) and/or your local agricultural extension agency office. If you feed a commercial brand name grain mix, contact the manufacturer for information about nutrients not listed on the guaranteed analysis on the bag.

If you are feeding a complete feed, you will be even more in the dark concerning exact levels of specific nutrients in your feed. A rough idea can be obtained by locating the roughage type on the label, as above, and matching this to a hay and grain diet chart or using the information in the complete feeds section of Basic Nutrition chapter to see how other ingredients will change the levels of nutrients. For truly accurate information, however, you will have to rely on information from the guaranteed analysis on the bag and any other details you can get from the manufacturer.

Many manufacturers of complete feeds refuse to release information about any nutrients not listed in the guaranteed analysis. If you really want to know what's inside that bag, you will have to have the feed analyzed.

HAY AND GRAIN ANALYSIS

To find out exactly what the vitamin and mineral levels are in your hay or grain, you will need to have it analyzed. Only a laboratory analysis can give you this precise information. Once you have the analysis, you can compare what you are feeding with what the horse actually needs. Both commercial sources (e.g., grain or supplement manufacturers) and academic sources (state universities, veterinary schools) are available for such tests. Shop around. Many places will give you a ration analysis that contains detailed suggestions for supplementation as well. Remember, however, that supplementation recommendations you receive will be based on the National Research Council's levels, unless otherwise specified. These may not be optimal for your horse and his activity level. (See Chapter 3, Nutrition A to Z).

FINE TUNING YOUR SUPPLEMENT NEEDS

In addition to soil factors affecting the level of nutrients in your diet, there are factors inside the horse that may make a diet insufficient, even if levels of vitamins and minerals seem to be adequate. For example, a horse with intestinal problems such as recurrent colic, diarrhea or heavy parasite load may not be receiving adequate B vitamins because the recommended levels of intake often rely heavily on the horse being able to absorb B vitamins that are manufactured inside his intestinal tract by the organisms that live there. For this to occur, everything must be functioning normally. Minerals may also compete with each other for absorption. This means that even though levels of a certain mineral in the diet are adequate, if another competing mineral is present in excess it may block the absorption (e.g., see Chapter 3, Nutrition A to Z entries for iron, copper, magnesium).

One way to get a better idea of what is going on, at least on the mineral front, is by hair mineral analysis. The mineral composition of hair reflects the mineral status of the body at the time the hair was growing. It can also provide clues into the horse's basic metabolism (fast or slow) and the level of free radical stress from exercise, drugs or environmental toxins, as well as if the horse has been exposed to toxic minerals.

Hair mineral analysis can tell you things about the horse's mineral status that blood tests cannot. This is because the body is designed to keep levels of all beneficial minerals in close control, even if tissue (muscle, heart, brain, bone, whatever) levels are not normal. In essence, the body's control mechanisms will read blood levels of substances and the body is designed to clear undesirable levels from the blood or to supplement low blood levels by pulling needed nutrients out of the tissues and into the blood. Blood levels may therefore be perfectly normal while tissues are either too high, too low or packed full of a toxic mineral that does not even appear in the blood.

Interpreting hair mineral analysis results takes considerable experience and knowledge. Deficiencies or excesses in the diet may appear, or you may see a pattern that does not reflect levels in the horse's diet. For example, a horse may be fed a diet that has a very high iron level and levels of copper and zinc that are only adequate. His hair mineral analysis, however, will show high levels of iron and low levels of copper and zinc—the absorption of the latter two being blocked by the high dietary iron. High hair mineral levels do not always mean excess in the diet. A horse with a high aluminum intake may also have high levels of calcium in the hair. This is not necessarily because the dietary calcium level is high. Aluminum displaces calcium in the bone and forces it out into the circulation. Other tissues, like hair, then take some of it up. The remainder is lost into the urine or the intestinal tract.

Hair mineral analysis is not a perfect test. Results and levels considered normal may vary between laboratories (just like they do for blood tests). As mentioned above, you cannot look at isolated mineral levels from hair mineral analysis and immediately assume that high levels are diet related. Too many factors intervene between the level of a mineral in the diet and the ultimate level in the hair—including chemical form of the mineral in the diet, dietary factors affecting absorption, digestive factors affecting absorption, increased need for the mineral in the body (e.g., growth, inflammation, fracture, pregnancy, etc.), disease and how well key organs are functioning (such as liver and kidney).

To understand hair mineral analysis and obtain useful information from it, it is necessary to understand how minerals interact with each other in the body and how specific patterns of low or high values and changes in the ratio of minerals to each other indicates there is a problem—either in the diet or within the horse itself. It is a complicated field, and interpretation should only be entrusted to laboratories where personnel have extensive background in clinical use of hair mineral analysis and those which routinely process equine samples.

The bottom line is that hair mineral analysis will show you which minerals are present in low, high or normal amounts. By understanding the interplay between minerals and how growth and health of the glandular system also affect mineral balance, the test may shed some light on whether it is a dietary imbalance or a factor within the horse itself (or both) that is causing the problem. If the remedy lies in changing the horse's intake of one or more minerals, it will be possible to design an individual supplement program that is highly specific.

As more is learned about how tissue mineral levels can reflect or influence abnormal conditions, hair mineral analysis may become a valuable tool in determining how best to approach treatment and in explaining the cause behind any failures of conventional treatments. For example, there is preliminary evidence that allergic problems such as hives or feed bumps may be closely tied to excessive iron intake and low tissue copper levels.

SPECIAL NEEDS

The above will give you basic supplement needs for your horse's diet and level of activity. The final step is to determine any special needs. If your horse has health problems, check Chapter 6, Relieving Health Problems Through Nutrition for those ailments that may respond to special supplementation levels. If your horse is at work, you will want to read Chapter 5, Nutrition and Performance to identify the type of basic diet and special supplements that may help improve performance.

PICKING A SUPPLEMENT

Once you have followed the steps above and established exactly what your supplement needs are, you are ready to choose a supplement. Equine supplements are a multimillion dollar business and the choices are staggering. Don't rely on advertising hype and what your neighbor says he or she uses to choose a supplement. Evaluate supplements armed with specific information about what you need and in what amounts. From there, choosing a supplement is a matter of matching your needs to what is in the product and determining which product is the most cost effective.

DECIPHERING LABELS

Product labels will contain information about ingredients and list such things as soybean meal, alfalfa, vitamin A, magnesium oxide and a sometimes seemingly endless cataloging of vitamins, minerals, herbs, preservatives and basic feedstuffs from such mundane things as corn to as exotic as bee pollen or mussels. Concentrate on the portion that gives an analysis of the product by vitamins and minerals.

Guaranteed analysis means that the product must live up to the amounts specified on the label. If it says 1 gram of calcium per scoop it must contain one gram of calcium per scoop. This means that not only was that amount put into the mix but that mixing and blending resulted in a product that was uniform. You won't find all that calcium sitting on the bottom of the can. Guaranteed analyses are the most reliable. Statements such as "average analysis" or "contains" do not promise each scoop will deliver the ingredients in the same amounts every time.

For labels to mean anything, you will have to be able to translate the analysis into the same units of measure you have for what you need. For example, you may know you need to feed an extra 60 mg per day of copper to your horse but the label lists copper in ppm (parts per million) or mg/kg (milligrams per kilogram). Many products also list the amount they contain per pound instead of per feeding. Read carefully. All of these are favorite techniques of manufacturers to make it look like there is more of a nutrient in there than there really is. For example, a product that contains 352 ppm of copper and is fed at a rate of 1 oz per day is only providing you with 10 mg of copper per day. You would have to feed six ounces of that supplement to get all the copper you need. The box below shows a price comparison example.

Price Comparison Example

	Product A	Product B
Copper	352 mg/oz	30 mg/oz
Cost @ package	$15@1kg	$35@4lb
Cost @ oz*	$.43@oz	$.547@oz
Cost @ serving size	6x.43=$2.58	2x.547=$1.09

*See table on page 220, "Converting Products to the Same Measure for Comparison"

Once you have located supplements that have all (or most) of the vitamins and minerals you are after, choose those that give a guaranteed analysis over those that do not. Last step is to determine real price to you on a daily basis. A serving is how much of the supplement you will have to feed each day. Let's return to the copper example. Product A contains 352 mg/oz of copper. Product B contains 30 mg/oz of copper. Product A costs $15.00 for a 1 kg container.

Product B costs $35.00 for a 4 pound container. The 1 kg container has 2.2 pounds in it for a cost per ounce of $15.00 divided by 2.2 = $6.81 per pound, divided by 16 = $0.43 per ounce. Product B costs $35.00 divided by 4 = $8.75 per pound, divided by 16 = $0.547 per ounce.

However, since you have to feed 6 oz of product A compared to only 2 oz of product B per day to give 60 mg/day, product B is by far the better buy.

FILLING IN THE GAPS

It is not at all unusual to find that you cannot meet all your supplement needs with one supplement. In this situation, begin by choosing a supplement that meets most of your needs then shop around to fill in the gaps. The product descriptions which follow this discussion can help you find supple-

Converting Products to the Same Measure for Comparison

The information on the products listed in this section has been transcribed directly from the labels on the products. This has resulted in some difficulty in comparing the products, because different manufacturers use different units of measure for the same ingredient. While all vitamins are measured in IU's (international units), other nutriments are measured variously in mg (milligrams), oz (ounces), lb (pounds), % (percent of serving size), or ppm (parts per million), and so on. If you want to compare the products on an equal footing, you will need to convert whatever measure appears on the label to mg's which is the most common unit. The table below will help you do that.

Convert from:	To mg
A. %	Convert to grams, ozs, or lbs by: (serving size in gram, ozs, or lbs) x (%) / 100 — then apply C, D, E
B. ppm	Convert to grams, ozs, or lbs by: (serving size in gram, ozs, or lbs) x ppm /1,000,000 — then apply C, D, E
C. grams	Multiply by 1,000
D. oz	Multiply by 28,400
E. lb	Multiply by 454,000
F. mcg	Multiply by .001
G. tablespoon	Multiply by 15,000
H. teaspoon	Multiply by 5,000

Examples:

A. If CA (calcium) is 10% and serving size is 2 lb, then to convert CA to lbs:
2 x 10/100 = .2 lbs. Then use formula E to convert from .2 lbs to mg.

B. If CA (calcium) is 150,000 ppm and serving size is 2 oz, then to convert CA to lbs:
2 x (150,000/1,000,000) = .3 oz. Then use formula D to convert from .3 oz to mg.

C. If CA (calcium) is listed as 2 grams, then to convert CA to mg:
2 x 1,000 = 2,000 mg

D. If CA (calcium) is listed as 2 ounces, then to convert CA to mg:
2 x 28,400 = 56,800 mg

E. If CA (calcium) is listed as .05 pounds, then to convert CA to mg:
.05 x 454,000 = 22,700 mg

F. If CA (calcium) is listed as 300,000 mcg, then to convert CA to mg:
300,000 x .001 = 300 mg

G. If CA (calcium) is listed as ˇ tbsp, then to convert CA to mg:
.25 x 15,000 = 3,750 mg

H. If CA (calcium) is listed as ° tsp, then to convert CA to mg:
.5 x 5,000 = 2,500 mg

ments that meet your needs. If you are not successful in getting everything you need from one or two supplements, it may be easier and more economical to make up the difference yourself using human products.

Every drug store, food store and health food store has a wide selection of individual vitamins and minerals. It is fine to use human vitamins or minerals for your horse. Follow the same instructions for deciphering the label. In general, you will find it much simpler to determine exactly how much of your nutrient is in a human product—it will be listed as amount per capsule or pill.

Preparing these products to give to the horse is a little more of a challenge. Pills need to be crushed finely for even mixing into the feed. An excellent way to do this is with an electric coffee grinder. These are inexpensive (about $20) devices available at hardware and department stores. Simply put the pills inside and push the button. Pills are ground to a fine powder within seconds.

ENSURING FRESHNESS

Vitamins and minerals, especially vitamins present in a mix with minerals where interactions can occur, will lose their potency over time. A few manufacturers list expiration dates on their products but this is the exception rather than the rule. Most products will have a lot and possibly also a batch number on the container. If you suspect the product is old, call the manufacturer and ask when that lot of product was manufactured and if it is still okay to use.

You will get the freshest products by ordering directly from the manufacturer. Next best choice is a large mail order supply catalog. Small local tack shops may be the worst place to buy your vitamins if they have passed through several distributors/middlemen in getting to you and/or have been on the shelf a long time.

CHOOSING YOUR BASIC SUPPLEMENT

Most people go with a multivitamin and mineral supplement to both fill their specific deficiency needs and to provide insurance against variations in quality of hay or grain. The most precise way to do this is to be armed with an analysis of your hay and grain as well as precise requirements for your horse's weight and use. However, you can do a reasonably accurate job of evaluating a supplement by using the charts in this book.

For example, let's follow a horse from weanling through adult using the diet/deficiency charts. We will look at how his supplement needs change with different ages and activity. The diet used will be a very common one—hay and oats. We will also assume the horse has access to and takes in enough salt. You can therefore ignore the red boxes pertaining to sodium and chloride (salt).

An unsupplemented hay and oats diet for a 4 month old weanling is deficient in crude protein, calcium, phosphorus, copper, zinc, selenium, iodine, vitamin D, vitamin E and lysine. Obviously this is the time to start using a high quality supplement that can meet these needs so that the baby gets off to a good start with his growth and development.

The 6 month old weanling fares no better on this diet and his levels of key trace minerals copper and zinc (both critical to the prevention of bone and joint development problems), selenium and iodine (both critical for normal production of thyroid hormone—essential for growth), vitamin D (the bone growth vitamin) and of vitamin E (important for normal muscles and immune system health) are not being met. Crude protein and lysine levels are also too low, which will hold back growth. Again, supplement is needed. This same deficiency pattern persists into the first year of life.

Once the horse becomes a year old, his vitamin D requirements drop somewhat as growth slows and this is now at an adequate level. However, all the other deficiencies listed above persist throughout this year on an unsupplemented oats and timothy diet.

Once the horse is a two year old, his supplement needs shift again. This is partially because his rate of growth has slowed but also because the relative amounts of grain and hay he should be receiving have changed—to less grain and more hay. The end result is that even though growth has slowed, the dropping of the amount of grain fed has actually resulted in more deficiencies.

The horse now has levels of calcium, phosphorus, magnesium, manganese, copper, selenium, iodine, vitamin E and lysine, which are all too low! The same holds true whether he is in training or not although training will mean he needs to eat a greater total amount of grain and hay.

By the age of 4, major demands of growth have been met and the horse's nutritional needs will settle into an adult pattern. For maintenance through moderate activity levels, he will be able to hold his weight well and get enough total protein (although not enough lysine) from a timothy and oats diet. However, energy and protein supplies will be marginal to deficient on this diet for heavy work loads, even if 50 percent of the diet is oats. In addition, intake of calcium, zinc, selenium, iodine, vitamin D and vitamin E will be borderline to frankly deficient at all levels of work and on varying intakes.

Once you have formulated your list of problem nutrients using the charts as described above, begin by looking at the major minerals—calcium, phosphorus, magnesium. If you feed alfalfa hay, you need a product that has a calcium:phosphorus ratio of less than 1—that is, more phosphorus than calcium in the product. High magnesium is desirable for diets based on alfalfa hay, but you will probably have some difficulty finding a multivitamin supplement containing high magnesium levels and may need to supplement magnesium separately if you have a special need (see tying-up). Pure magnesium supplements can be difficult to find but the search is worth it. If you feed a mixed hay or grass hay, look for a calcium:phosphorus ratio of greater than 1. Ideal calcium:magnesium will be 2:1.

Potassium is not an issue in supplements—your horse's hay will provide all the potassium he needs. Sodium and chloride needs are met by providing your horse with a salt block. It is best to use a plain, white salt block since the trace mineral blocks are formulated for other types of livestock, and the best

way to get trace minerals into your horse is by using an appropriate supplement in the feed.

Next in line are the trace minerals—iron, sulfur, manganese, zinc, copper, selenium, iodine and cobalt. You will NEVER need to supplement iron but it's going to be in there anyway. Choose a product that has as little iron as possible. Cobalt is another mineral you do not have to worry about as far as we know. Sulfur in its mineral form is probably of little use to the horse. More important will be any sulfur containing amino acids (methionine). This leaves manganese, zinc, copper, selenium and iodine—any or all of which may be deficient in your diet. Using the basic nutrition charts (Chapter 1, Basic Nutrition) as a guide, identify the deficient elements. If in red, go to Chapter 3, Nutrition A to Z to see what the recommended daily supplementation is and make sure the product will give you that amount.

Vitamins are next in line. For all the B vitamins, follow recommendations in A to Z; same for the fat soluble vitamins A, D and E. You will find that many manufacturers use quite a bit of vitamin A and are short on E. Vitamin C may also be absent or in a very small amount. You may need to go with single ingredient products or an antioxidant product to make up those differences. Vitamin K (menandione) is added to many multi products but is of no value. Some multi products also contain important amino acids—lysine and methionine. Make sure they are more than window dressing ingredients. Look for methionine at a level of 1.5 to 2.0 grams per daily serving; lysine 2 to 4 grams per daily serving.

Once you start reading labels and evaluating before you buy, you will be amazed at the differences between products. Big names and big advertising budgets do not necessarily mean you are getting the best product for your needs. Buying the incorrect product is not only an expensive mistake, it could also mean you will not get maximum benefit or, worse yet, create imbalances.

ADEQUACY OF TIMOTHY & OATS DIET AT VARIOUS AGES

Note: Proportions of hay and grain will change depending on the age of the animal. Refer to Chapter 2, Mares and Growing Horses, for further details.

LEGEND

Weanling 4 Months · Weanling 6 Months · Yearling · 18 Months No Training · 24 Months No Training · 24 Months Trained

DEFICIENCY ANALYSIS

- (green) adequate
- (yellow) marginal to inadequate
- (red) inadequate
- (blank) level of this nutrient unknown

Diet	DE	CP	Ca	Ph	Mg	K	Na	Cl	Su	Fe	Mn	Cu	Z	Se	I	Co	VitA	VitD	VitE	Thia	Ribo	Lys
Weanling 4 Months																						
Weanling 6 Months																						
Yearling																						
18 Months No Training																						
24 Months No Training																						
24 Months Trained																						

DE digestible energy   Ph phosphorus   Na sodium   Fe iron   Z zinc   Co cobalt   VitE vitamin E   Ribo riboflavin
CP crude protein   Mg magnesium   Cl chloride   Mn manganese   Se selenium   VitA vitamin A   Thia thiamine   Lys lysine
Ca calcium   K potassium   Su sulfur   Cu copper   I iodine   VitD vitamin D

GLOSSARY

BRITTLE FEET: Hooves prone to cracking, splitting and peeling.

CARBOHYDRATE LOADING: A process developed for human weight lifters and runners to increase the supply of muscle glycogen.

CHOKE: Food caught in the esophagus.

CREEP: An adjective used to describe the process of feeding the foal grain before it is weaned.

CURB: Swelling of ligament just below the point of the hock.

DESMITIS: Lameness. Heat, swelling and pain of a ligament such as the suspensory or curb.

ELECTROLYTES: Minerals which exist in the blood or cells in an electrically charged form.

ENDOTOXEMIA: Potentially fatal body wide illness resulting from absorption of toxins produced by bacteria in the intestine.

EPIPHYTIS: An inflammation of the growth plate, the region where bone grows.

ESTERIFIED VITAMIN C: A form of vitamin C where the natural high acidity has been reduced.

FEED BUMPS: Hives believed to be caused by type of feed.

HARD KEEPERS: Horses that have trouble gaining and holding weight.

HAY BELLY: Swollen, rounded appearance to the lower abdomen; seen in horses eating large amounts of hay and/or with poor digestion.

HEEL SCRATCHES: Cracked open skin on back of heels and back of pastern.

HYPOTHYROIDISM: Abnormally low levels of thyroid hormone.

LACTATE ACCUMULATION: Buildup of lactic acid in the blood or tissues.

LAMINITIS: An inflammation of the live tissue inside the hoof. Also known as founder.

LASIX: A drug (furosemide) which helps most horses with lung bleeding. This diuretic also decreases pressure on the right side of the heart.

LUNG BLEEDING: Symptoms include decreased exercise tolerance, coughing after work, blood seen at nostrils or during exam of lungs with endoscope.

METABOLIC ACIDOSIS: Body-wide condition where the pH of the blood is more acid than normal.

METABOLITES: Substances produced when another substance is broken down.

MILK TETANY: Muscle weakness and hyperexcitability caused by sudden drop in blood and tissue calcium when mare starts to make milk.

NATIONAL RESEARCH COUNCIL: Federal government agency that sets nutritional guidelines.

OSTEOCHONDROSIS DESSICANS: A congenital disease that results in improper formation of the joint cartilage.

POLYSACCHARIDE STORAGE DISEASE: A muscle defect that can cause tying-up.

STOCKING UP: Swelling of the lower legs.

THUMPS: A condition where the diaphragm (the muscle that moves the chest up and down when the horse breathes) contracts suddenly and violently with each heartbeat, causing a visible jerking/thumping behind the horse's ribs.

TYING-UP: Exercise induced muscle spasm, pain on palpitation of muscles, reluctance to move.

VITAMIN A (BETA-CAROTENE): The naturally present form in plants that is composed of two vitamin A molecules hooked together.

VITAMIN A: Retinyl palmitate or other retinyls.

VITAMIN B1: Thiamine.

VITAMIN B2: Riboflavin.

VITAMIN B3: Niacinamide or niacin.

VITAMIN B5: D-Pantothenate or pantothenic acid.

VITAMIN B6: Pyriodoxine HCL.

VITAMIN B12: Cyanocobalamine.

VITAMIN C: Ascorbic acid or various ascorbates.

VITAMIN D3: Cholecalciferol.

VITAMIN E: D-alpha-tocopherol succinate.

VITAMIN K: Phytonadione or menandione.

WHITE MUSCLE DISEASE: Potentially fatal disease of skeletal and heart muscle caused by selenium deficiency.

INDEX

Note: Bold page numbers refer to major discussions of nutrients and diseases.

Abdomen, 57, 115, 127, 168, 185
Acetylcholine, 157
Acid rain, 19, 21, 22
Adrenal glands, 25, 112
Aggression, **178-79**
Aging, 47, 87, 95, 100, 105, 106, 119, 120, 131, 166, **204-5**
Alanine, 124, 125, 128, 129, 142, 143, 153, 160
Alfalfa and amount of feed, 146
  and basic diet, 1
  and bloating/gas, 180
  and choosing basic supplement, 221
  and complete feeds, 10
  deficiency of, 5, 7
  diet discussion of, 5, 7
  and growing horses, 25
  and mineral levels, 18
  in mixed hays, 9
  oats mixed with, 7
  for pregnant/lactating mares, 34, 35, 36
  and protein-sweet feed, 17
  and racing horses, 147
  in stock feed, 11
  and sweet feeds, 15, 17
  for two year olds, 32, 33
  and tying-up/muscle pain, 211
  and use requirements, 146
  for weanlings, 29
  and weight, 202, 203
  for yearlings, 30
  see also specific nutrient
All Stock Feed. See Stock feed
Allergies, 99, 120, 121, 134, 135, 166, 189, 196, 201
Aluminum, 19, 21, 218
Amino acids, **153**
  branched chain, 98, 117, 124-25, 128-29, 142-43, **153**, 211
  and choosing basic supplement, 221
  and consequences of inadequate nutrition, 23
  deficiencies of, 59, 109, 125, 129, 133, 143
  essential and nonessential, 109,

124, 128, 132, 142
  functions of, 25, 41, 59, 109, 110, 128, 132, 142, 143, 153, 211
  in growing/pregnant horses, 25
  and growth, 59, 109, 125, 129, 133, 143
  indications for needed supplementation of, 124, 129, 143
  interactions with, 59, 81, 107, 109, 110, 124, 125, 129, 133, 143
  method/timing of, 153
  and performance, 59, 153
  sources of, 58, 128, 132, 142
  supplementation of, 128, 132, 142
  toxicity of, 59, 129
  see also specific amino acid or function
Aminoglycocide antibiotics, 59
Anabolic steroids, 116, 117, 126, 198
Anemia, 46, 53, 93
Antioxidants, deficiencies of, 47, 77, 87, 91, 95, 100, 106, 130, 131
  functions of, 47, 76-77, 86-87, 94-95, 105, 130, 176, 195, 204, 205, 206, 211, 212, 213
  indications for needed supplementation of, 47, 77, 87, 95, 100, 106, 130, 131
  interactions with, 47, 77, 85, 87, 91, 95, 99, 100, 106, 120, 130, 131
  supplementation of, 77, 91, 95, 99-100, 106, 130, 131
  toxicity of, 87, 91, 100, 106, 130, 131
  see also specific antioxidant or function
Appetite and constipation, 186
  and iodine, 55
  and pneumonia, 208
  and sodium chloride, 79
  and vitamin B, 49, 53, 65, 66, 73, 75, 82, 150
  and vitamin D, 89

see also Weight
Arabian horses, 156
Arginine, 59. See also L-arginine
Arthritis, **176-77**
  and antioxidants, 176
  and bioflavinoids, 176
  causes of, 177
  and chondroitin, 101, 102, 176
  and copper, 47, 176
  and DMG, 111, 162
  and essential fatty acids, 112, 113
  and fat, 114, 115
  and glucosamine, 119, 176
  and grape seed extract, 120, 121, 166
  and iron, 57
  and manganese, 63, 176
  pathology of, 177
  and perna mussel, 136, 176
  and phenylalanine, 137
  problem description of, 176
  and SOD, 140, 141
  supplement program for, 176
  symptoms of, 176
  and TMG, 162
  treatment notes for, 176
  and vitamin C, 172, 176
  and work level, 177
  and zinc, 176
ATP, 69, 122, 146, 147, 160, 167

Bahia grass, 43
BAP (beta-aminopropionitrill), 182
Barley, 11, 62
Basic nutrition for competition, 147
  and complete feeds, 10, 12-13
  diets for, 1-17
  for endurance, 147-48
  for light work, 148
  and mineral levels, 18-22
  for moderate work, 148
  for racing horses, 147
  and stock feeds, 11
  and sweet feeds, 10, 14-16, 17
  for training, 147
BCAA. See Amino acids: branched chain; specific amino acid
Beet pulp, 147, 205, 217

Behavioral problems, **178-79**
Bermuda grass, 43
Beta-aminopropionitrill (BAP), 182
Beta-carotene, 84, 85, 197
Beta-hydroxy-beta-methylbutyrate
　(HMB), **98**, 153
Betaine. *See* TMG
Bicarbonate, **151**, 159, 164, 171
Bioflavinoids, **99-100**, **152**
　and arthritis, 176
　and bowed tendon/tendonitis, 181
　and brittle/cracking feet, 183
　and desmitis/suspensories/curbs,
　　187
　and energy, 99
　and exercise, 99, 100
　and founder/laminitis, 193
　and grape seed extract, 120, 166
　and heaves/COPD, 196
　and immune system, 99
　and infections, 99, 100, 152, 172,
　　209, 213
　and lung bleeding, 172, 200
　and muscles, 99
　and OCD, 206
　and pneumonia, 208
　and vitamin C, 87, 99, 100, 152,
　　172, 176
　see also Antioxidants
Biotin, **40-41**, 150, 181, 183, 187,
　191, 193, 197, 202, 209
Birth defects, 67
Blindness, 77
Bloating, 107, 170, **180**
Blood cells red, 48, 53, 56, 57, 59,
　93
　white, 56, 91
Blood clotting, 42, 81, 92, 93
Blood pressure, 137, 163
Blood sugar, 124, 125, 128, 129,
　147, 154, 158
Blood vessels, 47, 87, 95, 99, 100,
　106, 120, 131, 166
Body temperature, 79
Bones and joints and alfalfa, 5
　and calcium, 25, 42
　and chondroitin sulfate, 101, 102
　consequences of inadequate
　　nutrition for, 23, 25
　and copper, 19, 46, 47
　development of, 63
　and DMG, 111
　and fluorine, 50, 51
　in foals, 25
　and gamma oryzanol, 117

　and glucosamine, 118, 119
　in growing horses, 23, 25, 26
　and manganese, 63
　in mares, 26
　and methionine, 132
　and MSM, 134, 135
　and perna mussel, 136
　and phosphorus, 69
　and protein, 25, 146
　and sulfur, 81
　and vitamin A, 85
　and vitamin C, 87, 172
　and vitamin D, 26, 89
　in weanlings, 28
　and zinc, 95
　see also specific disease
Bowed tendons, **181-82**
Brain, 47, 73, 105, 112. *See also*
　Seizures
Bran mash, 186
Brans and diarrhea, 189
　minerals in, 46, 60, 68, 80
　vitamin B in, 48, 52, 64, 66, 72,
　　74, 82, 150
　see also type of bran
Breathing, 77, 87, 120, 185, 196,
　208. *See also* Respiratory
　system
Breeding, 85, 116, 117
Brewer's grains, 76, 205
Brittle feet, **183-84**
Brome grass, 43
Burns, 109, 110

Calcium, **26**, **42-43**
　and acid rain, 19
　and aging, 204
　and alfalfa, 5, 57
　and behavioral problems, 178,
　　179
　in brans, 68
　and choosing basic supplement,
　　220
　and complete feeds, 10
　and cramping, 43
　deficiency of, 43
　and determining supplement
　　needs, 218
　and diarrhea, 189
　and electrolytes, 164
　and endurance, 43
　functions of, 42
　in grains, 42
　in grasses, 42, 43
　for growing horses, 26, 43

　and guaranteed analysis, 10
　in hays, 9, 42, 43
　indications for needed supplement
　　of, 43
　interactions with, 43, 57
　and Lasix, 163
　and magnesium, 43, 61, 179, 189
　for mares, 26, 34, 43
　and methionine, 133
　and mineral levels, 18, 19
　and mixed hay diet, 9
　and muscles, 42, 43, 211
　and phosphorus, 26, 43, 69
　and potassium, 71
　and protein, 43
　and sodium phosphate, 171
　sources of, 42
　supplementation of, 42, 43
　in sweet feeds, 15
　and timothy, 4
　toxicity of, 43
　and vitamin B, 83
　and vitamin D, 43, 88, 89
　and vitamin K, 93
Calories, **25**
　and carbohydrate loading, 154
　and chromium, 103
　and epiphysitis, 192
　and fat, 114, 115
　and growing horses, 25
　and guaranteed analysis, 10
　for lactating mares, 35, 36
　and protein, 146
　for two year olds, 32
　and use requirements, 146, 147
　for weanlings, 29
　and weight, 203, 214
　for yearlings, 30, 31
　see also Energy; *specific type of*
　　diet
Cancer, 47, 87, 95, 100, 106, 120,
　131
Canola oil, 91
Carbohydrates, **145**, **154-55**
　and behavioral problems, 178
　and carnitine, 156
　and chromium, 104
　and creatine, 160
　and endurance, 145, 147, 148,
　　154-55
　and energy, 147
　and exercise, 145, 155
　and fat, 165
　functions of, 145, 147
　and grains, 147, 154

and grass, 145
in growing horses, 26
and hays, 145, 154
loading of, 148, **154-55**
in mares, 26
and metabolism, 147
and muscles, 145, 154, 155
and performance, 145, 154-55
and phosphorus, 69
and special diets, 147
and sulfur, 81
for three-day event horse, 148
and tying-up/muscle pain, 211, 212
and vitamin B, 73, 83, 150
and zinc, 95
see also Glycogen
Carnitine, 98, **156**, 211
Carotene, 84, 85
Carrots, 85, 203
Cartilage. *See* Bones and joints
Catechin, 99
Cattle, 20
Cattle feed, 11
Cell division, 53, 63, 65, 83
Chickens, 19
Chloride, **78-79**, 164, 220. *See also* Glucosamine hydrochloride
Cholesterol, 63, 65, 98
Choline, **157**, 202, 212
Chondroitin sulfate, **101-2**, 118, 119, 136, 175, 176, 206
Chromium, **103-4**, **158**
Citrate, **159**, 164
Citric acid, 147
Citrus extracts, 100
Clover hay, 93
Coat, 57, 59, 85, 113, 114, 115, 191, 210
Cobalt, 18, **44-45**, 204, 221. *See also* Vitamin B12
Coconut oil, 91, 214
Coenzyme Q10, **105-6**, 195, 196, 200, 204, 208, 212
Colic, 77, 81, 139, 179, 180, **185**, 217
Collagen, 81, 134
Competition, 13, 61, 114, 147, 150, 165, 211, 212. *See also* Performance
Complete feeds, 10, 12-13, 46, 217
Connective tissues and amino acids, 41
and chondroitin sulfate, 101
and copper, 46, 176, 181

and glucosamine, 118
in growing horses, 26
in mares, 26
and MSM, 134, 135
and sulfur, 81
and vitamin C, 26, 87
Constipation, **186**
Convulsions, 73
Coordination, 73
COPD, **196**
Copper, **19**, **46-47**
and aging, 204, 205
and bones and joints, 19, 46, 176, 181, 187, 206, 211
and brans, 46
and brittle/cracking feet, 183
and chondroitin sulfate, 102
and complete feeds, 10, 46
and connective tissues, 176, 181
and cystine, 110
deficiency of, 19, 46, 47
and determining/selecting supplement needs, 218, 221
and exercise, 47
and foals, 46
and founder/laminitis, 193
functions of, 18, 46-47
and gingivitis, 195
and glucosamine hydrochloride, 119
in growing horses, 26
and hays, 46
and immune system, 47
indications for needed supplementation of, 47
and infections, 47, 209
and infertility, 198
interactions with, 47
and iron, 46, 57
in mares, 26, 34, 46, 47
and mineral levels, 18, 20
and MSM, 135
and muscles, 47
NRC levels for, 26, 47
and perna mussel, 136
and SOD, 140
sources of, 46
in stock feed, 11
and sulfur, 81
supplementation of, 46, 47
and sweet feeds, 46
toxicity of, 47
in two year olds, 32
and vitamin C, 87
for weanlings, 28, 29

for yearlings, 30, 31
and zinc, 47, 95
Corn, 11, 31, 32, 33, 56, 59, 62, 74, 132, 154, 205
Corn syrup, 107
Corticosteroids, 176, 207
Cortisol, 87, 104
Coughing, 196, 200, 208, 213
Cracking feet, **183-84**
Cramping, 26, 43, 55, 61, 71, 73, 82, 106, 131, 160
Creatine, **107-8**, **160-61**
Creatine phosphate (CrP), 147, 160, 161
Creep feeding, 36
Curbs, **187-88**
Cyanocobalamin. *See* Vitamin B12
Cysteine, 81, 132
Cystine, 81, **109-10**, 133

Dehydration/hydration, 79, 163, 189, 190
Depression, 127, 168
Desmitis, **187-88**
Diabetes, 129, 131
Diaphragm, 43
Diarrhea, 49, 61, 65, 77, 127, 168, 179, 180, **189-90**, 217
Diet basic, 1-9
comparison of human and horse, 37-38
and determining supplement needs, 217
evaluation of, 1
special, 147-48
see also Complete feeds; Sweet feeds; *specific diet*
Digestive tract and alfalfa diet, 5
and alfalfa-oats diet, 7
and chondroitin sulfate, 102
and gas, 180
and gingivitis, 195
and indigestion, **185**
and mixed hay-oats diet, 8
and perna mussel, 136
and probiotics, 138, 139
and protein, 170
and SOD, 140
and timothy-oats diet, 6
and vitamin A, 85
and vitamin B, 64, 65, 73, 82, 150
and vitamin K, 93
see also Weight
Dimethylglycine (DMG), 98, **111**, **162**, 212

Diuretics, 163
DMSO, 134
Dopamine, 137
Dry skin, **191**

Ears, 112
Electrolytes, 43, **163-64**, 189, 200.
   See also specific electrolyte
Emphysema, 208
Endorphins, 137
Endotoxemia, 154
Endurance and amino acids, 125,
   129, 143, 153
   basic diet for, 2, 147-48
   and blood sugar, 147
   and calories, 147
   and electrolytes, 163, 164
   and energy, 147
   and grains, 147
   and hays, 147
   and muscles, 147, 148
   for three-day event horse, 148
   and training, 147
   use requirements for, 146
   and weight, 147
   see also specific nutrient
Energy, **25**
   and amino acids, 59, 110, 124,
      125, 128, 129, 133, 142, 143
   and basic diet, 2
   and endurance, 147
   and epiphysitis, 192
   fuels for, 145-46
   for three-day event horse, 148
   and tying-up/muscle pain, 212
   see also Calories; *specific*
      nutrient
Enzymes and amino acids, 153
   and arthritis, 176
   and carnitine, 156
   and fat, 114-15, 145, 165
   and glucosamine, 118
   and magnesium, 60
   and MSM, 134
   and phenylalanine, 137
   and tying-up/muscle pain, 211
   and zinc, 95
Epinephrine, 87, 137
Epiphysitis, 25, 117, **192**
Equipoise, 116
Essential fatty acids, **112-13**, 114,
   115, 145, 178, 202
Excitability, 43, 82, 150, 180. *See*
   also Irritability
Exercise and amino acids, 124, 125,

128, 129, 142, 143, 153
   and antioxidants, 131
   and behavioral problems, 178,
      179
   and bowed tendon/tendonitis, 182
   and desmitis/suspensories/curbs,
      188
   and electrolytes, 164
   and epiphysitis, 192
   and founder/laminitis, 194
   fuels for, 145-46
   and heaves/COPD, 196
   and lung bleeding, 200, 201
   and pneumonia, 208
   and tying-up/muscle pain, 211
   and weight, 202, 203
   see also specific nutrient
Extruded feeds, 205
Eyes, 77, 85, 112

Face, 69
Fat, **26**, **114-15**, **145**, **165**
   and alfalfa-oats diet, 7
   and basic diet, 147
   and behavioral problems, 178
   and bloating/gas, 180
   and carbohydrates, 154, 155, 165
   and carnitine, 156
   and chromium, 103, 158
   and endurance, 114, 115, 147, 165
   and energy, 115, 165
   and exercise, 145
   functions of, 145
   and grains, 114, 165
   and growing horses, 26
   and hays, 115, 165
   and iodine, 55
   and L-arginine, 168
   of lactating mares, 35
   and manganese, 63
   and mixed hay-oats diet, 8
   and muscles, 114-15, 145, 165
   and performance, 145, 146, 165
   for three-day event horse, 148
   and timothy-oats diet, 6
   and training, 147
   and tying-up/muscle pain, 211
   and use requirements, 146
   and vitamin A, 85
   and vitamin B, 67, 83
   and vitamin C, 172
   and vitamin E, 91
   and weight, 165, 202, 214
Fat soluble vitamins. *See* Vitamin A;
   Vitamin D; Vitamin E; Vitamin K

Fatigue and amino acids, 124, 128,
   142
   and bioflavinoids, 99
   and coenzyme Q$_{10}$, 105
   and copper, 47
   and iron, 57
   and lipoic acid, 130
   and niacin, 65
   and selenium, 77
   and vitamin C, 87
   and zinc, 95
Fatty acids, 26, 65, 112-13, 136,
   147, 156, 158, 202. *See also*
      Essential fatty acids
Feeds and constipation, 186
   determining amount of, 146
   and electrolytes, 164
   extruded, 205
   and fat, 165
   how to select, 12
   mineral levels in, 18-22
   price of, 11
   and use requirements, 146
   and weight, 214-215
   see also specific type of feed
Feet, 95, **183-84**, 193. *See also*
   Hooves
Fertility, 19, 55, 57, 59, 85, 95, 116,
   117, **198-99**
Fertilizers, 19, 21
Fever, 208, 213
Fiber, 10, 139
Flatulence, 133, 138, **180**
Flax oil, 91, 115, 178, 181, 183,
   187, 191, 193, 197, 205,
   206, 209
Fluorine, **50-51**
Foals, 19, 25, 26, 34, 46, 57, 89,
   138, 189
Folic acid, 48, **52-53**, 67, 75, 150,
   178, 202, 204, 212
Founder, **193-94**
Fractures, 89, 128
Free radicals and bioflavinoids, 99,
   172
   and coenzyme Q$_{10}$, 105
   and copper, 47
   functions of, 47, 76-77, 86-87,
      94-95, 172, 204, 205
   and lipoic acid, 130
   production of, 47, 77, 87, 95, 172
   and selenium, 76-77, 91
   and SOD, 140
   and vitamin C, 86-87, 172
   and vitamin E, 91

and zinc, 94-95
Fungal infections, 208, 209

GAGs (glyco-saminoglycans), 118, 119, 136
Gamma oryzanol, **116-17**, 127, 168, 198, 205
Gas, excessive, 133, 138, 178, **180**
Genetics, 167, 212
Gingivitis. *See* Gum disease
Glucosamine, 102, 119, 134, 175, 176, 195, 197, 206, 210
Glucosamine hydrochloride, **118-19**
Glucosamine sulfate, 118-19
Glucose blood, 103, 131, 158, 161
    and carbohydrate loading, 154, 155
    and chromium, 103, 158
    and endurance, 147
    and HMB, 98
    and lipoic acid, 130-131
    and pantothenic acid, 66
    and vitamin B, 40
Glutamine, 125, 129, 143
Glutathione, 77, 111
Glycine, 127, 160, 168
Glycogen and amino acids, 142, 153
    BCAAs as substitute for, 124, 125, 128, 129, 142, 143, 153
    deficiency of, 79
    and endurance, 148
    and endurance/performance, 147, 148, 153
    as energy source, 73, 145, 154
    and exercise, 124, 125, 128, 129, 142, 143, 153
    and fat, 115, 145, 165
    and sodium chloride, 79
    and sodium phosphate, 171
    and vitamin B, 73
Glycosaminoglycans, 118-119, 136
Goiter, 55
Grains and aging, 205
    and amount of feed, 146
    analysis of, 217, 220
    and basic diet, 2
    and behavioral problems, 178
    and bloating/gas, 180
    and determining supplement needs, 217
    and diarrhea, 189
    and endurance, 147
    function of, 6
    in lactating mare diets, 35, 36
    in pregnant mare diet, 34

and race horses, 212
    salt in, 79
    in stock feed, 11
    and sweet feeds, 10, 15
    in two year old diet, 32, 33
    and tying-up/muscle pain, 212
    and use requirements, 146
    for weanlings, 28, 29
    and weight, 203, 214
    for yearlings, 30, 31
    see also specific nutrient or grain
Grape seed extract, **120-21**, 152, **166**, 196, 200, 208, 213
Grass and amount of feed, 146
    and basic diet, 3
    and choosing basic supplement, 221
    and complete feeds, 10
    and constipation, 186
    and racing horses, 147
    in sweet feeds, 15
    and use requirements, 146
    and weight, 202, 203
    see also Pasture; *specific nutrient*
Growing horses bones and joints in, 23, 25, 26
    and calcium, 43
    and chromium, 104
    consequences of inadequate nutrition in, 23, 25
    and gamma oryzanol, 116, 117
    immune system in, 25
    infections in, 25
    and manganese, 63
    minerals for, 26-27
    nutrient requirements and balancing rations for, 24
    and phosphorus, 69
    and protein, 145
    role of major nutrients for, 25-27
    sample diets for, 27-36
    and stress, 25
    vitamins for, 26-27, 85, 89, 91
Growth and amino acids, 59, 109, 125, 129, 133, 143
    and epiphysitis, 192
    and iodine, 54
    and L-arginine, 126, 127, 168
    and protein, 145
    and vitamin B, 40
    see also Growing horses
Gruels, 205
Guaranteed analysis, 10, 217, 219
Gum disease, 105, 106, 121, 166, **195**

Hair, 21, 55, 77, 85, 109, 111, 145, 191, 209, 218
Hay belly, 138
Hays and amount of feed, 146
    analysis of, 217, 221
    and basic diet, 1, 2
    and choosing basic supplement, 221
    and determining supplement needs, 217
    and diarrhea, 189
    and electrolytes, 163
    and endurance, 147
    inadequacy of, 2
    in lactating mare diets, 35, 36
    minerals in, 1-2
    in pregnant mare diet, 34
    and racing horses, 147
    in sweet feed, 17
    in two year old diet, 32, 33
    and use requirements, 146
    vitamins in, 1-2
    for weanling diet, 28, 29
    and weight, 202, 203
    for yearlings, 30, 31
    see also Mixed hays; *specific nutrient or type of hay*
Headaches, 127
Healing and bioflavinoids, 100, 152
    and bowed tendon/tendonitis, 182
    and calcium, 89
    and coenzyme Q_{10}, 106
    and copper, 47
    and cystine, 109
    and desmitis/suspensories/curbs, 188
    and glucosamine, 118
    and leucine, 128
    and lipoic acid, 131
    and lysine, 59
    and perna mussel, 136
    and viral respiratory infections, 213
    and vitamin C, 87
Heart and bioflavinoids, 100
    and calcium, 42, 43
    and coenzyme Q_{10}, 106
    and copper, 47
    and diuretics, 163
    and electrolytes, 163
    and iron, 57
    and lipoic acid, 131
    and magnesium, 60
    and potassium, 70, 71
    viral infections in, 213

and vitamin A, 85
and vitamin C, 87
and zinc, 95
Heaves, **196**, 208
Heavy work and alfalfa-oats diet, 7
and arthritis, 177
and basic diet, 2
and chondroitin sulfate, 101
and complete feeds, 13
and electrolytes, 163
and fat, 114
and gamma oryzanol, 116
and glucosamine hydrochloride,
118
and grape seed extract, 166
and HMB, 98
and iodine, 55
and lipoic acid, 130
and lung bleeding, 200
and mixed hay-oats diet, 8
and perna mussel, 136
and phosphorus, 69
and sodium chloride, 79
and timothy hay-oats diet, 6
use requirements for, 146
and vitamin B, 49, 65, 67, 73, 75,
83
and vitamin D, 89
and weight, 214
Heel scratches, **197**
Hemoglobin synthesis, 124
Hemp oil, 115, 178, 181, 183, 187,
191, 193, 197, 206, 209
Hesperidin, 99, 152, 196, 200, 208
Histamine, 99, 120, 166
Histidine, 59, 109, 124, 128, 132,
142
HMB (beta-hydroxy-beta-
methylbutyrate), **98**, 153
Honey, 107, 160
Hooves biotin for, 40, 41
brittle/cracking, 183-84
and copper, 47
and DMG, 111
and essential fatty acids, 112, 113
and founder/laminitis, 193
and lysine, 41, 59
and methionine, 41, 132, 133
and MSM, 134, 135
and protein, 145
and selenium, 77
separation of, 77
and sulfur, 81
and vitamin B, 41, 150
and zinc, 41, 175

Hormones and aging, 204
and behavioral problems, 178
and calcium, 26, 42
and carbohydrate loading, 154
and fat, 114
and fatty acids, 26
and growing horses, 26
and HMB, 98
and infertility, 198
in mares, 26, 179
and MSM, 134
and vitamin C, 172
and weight, 203
Horses, human diets compared with
those of, 37-38
Human vitamins/minerals, 219-20
Hyaluronic acid, 101, 118, 176, 206
Hydrogen ion concentrations, **21**
Hypersensitivity, 61
Hyperthyroidism, 215
Hypoglycemia, 125
Hypothyroidism, 55, 77, 95
Hypoxanthine riboside, 167. *See*
Inosine

Immune system and consequences
of inadequate nutrition, 25
and gingivitis, 195
in growing horses, 25, 26
in mares, 26
minerals for, 26
and skin infections, 209, 210
and viral respiratory infections,
213
see also specific nutrient
Impaction, **186**
Indigestion, **185**
Infections, **209-10**
and antioxidants, 131
in growing horses, 25
and heel scratches, 197
and infertility, 198
and probiotics, 138
*see also specific nutrient or type
of infection*
Infertility. *See* Fertility
Inflammation and bioflavinoids, 99,
152
and chondroitin sulfate, 101
and coenzyme Q_{10}, 105
and copper, 47
and DMG, 111
and essential fatty acids, 112, 113
and fat, 114
and glucosamine hydrochloride, 119

and grape seed extract, 120, 121,
166
and lipoic acid, 130
and perna mussel, 136
and selenium, 77
and SOD, 141
and vitamin C, 86-87, 172
and zinc, 95, 174
*see also specific type of inflam-
mation*
Injuries, 131, 135. *See also* Sprains;
Strains; Wounds
Inosine, **122-23**, **167**
Inositol, 212
Insulin, 103, 104, 107, 131, 154,
158, 161
Intestinal tract and calcium, 93
and carbohydrates, 145
and cobalt, 44
and creatine, 107
and MSM, 134, 135
parasites in, 185, 186
and probiotics, 138-139, 214
and protein, 170
ulcers in, 134
and vitamin A, 84
and vitamin B, 48, 49, 52, 53, 64,
65, 66, 72, 73, 74,
75, 82
and vitamin K, 92, 93
and weight, 214
Intravenous fluid nutrition, 49, 150
Iodine, **26**, **54-55**, 184, 197, 210,
221
Iron, 26, 46, **56-57**, 63, 87, 91, 138,
182, 214, 218, 222
Irritability, 61, 65, 73, 82, 127, 137,
150, 168. *See also* Excitability
Isoleucine, 59, 109, **124-25**, 128,
132, 142, 153, 211
Ivermectin, 214

Jaw, 69
Joint problems. *See* bones and
joints; *specific disease*

Kentucky bluegrass, 43
Keratan, 111
Kidneys, 51, 57, 59, 89, 93, 146,
211
Krebs cycle, 147

L-arginine, **126-27**, **168**
L-lysine, 213
L-ornithine alpha-ketoglutarate,

127, 168

Labels, 219-220. *See also* Guaranteed analysis

Lactate, 82, 98, 104, 124, 128, 142, 153, 156, 171

Lactating mares, 26, 35-36, 43, 69, 91

Lactic acid, 69, 105, 153

Lameness. *See* Legs

Laminitis, 93, 154, **193-94**

Lasix, 71, 79, 163, 164, 200

Last trimester, 34

Lawrence, L. A., 24

Legs, 25, 69, 181, 187, 206, 209

Leucine, 59, 98, 109, 124, **128-29**, 132, 142, 153, 211

Ligaments. *See* Tendons/ligaments

Light work, 2, 5, 9, 13, 14, 79, 146, 148

Limestone, 21

Linoleic/linolenic acid, **112-13**, 204

Linseed oil, 91, 114

Lipoic acid, 105, 106, **130-31**, **169**, 211

Liver, 57, 81

Lungs and bioflavinoids, 99, 152

  bleeding from, 79, 87, 93, 99, 120, 121, 152, 166, 172, **200-201**, 208

  and grape seed extract, 120, 121, 152, 166

  and heaves/COPD, 196

  infections in, 196

  pneumonia in, 208

  and sodium, 79

  and sulfur, 81

  and vitamin C, 87, 152, 172

  and vitamin K, 93

Lysine, **58-59**

  and aging, 204

  and alfalfa, 58

  and bowed tendon/tendonitis, 181

  and brittle/cracking feet, 183

  and choosing basic supplement, 221

  and corn, 59

  deficiencies of, 59

  and desmitis/suspensories/curbs, 187

  as essential amino acid, 109, 124, 128, 132, 142

  and exercise, 59

  and founder/laminitis, 193

  function of, 59

  and grains, 58, 59

  and grass, 58, 59

  and growing horses, 25

  and hays, 58, 59

  and heel scratches, 197

  and hooves, 41

  indications for supplementation of, 59

  and infections, 209

  interactions with, 59

  in lactating mare diets, 35, 36

  and mixed hay diet, 9

  and muscles, 59

  for pregnant mares, 34

  and protein, 58

  and skin, 191, 209

  sources of, 58

  supplementation of, 58, 59

  and timothy, 4

  toxicity of, 59

  for weanlings, 28

Magnesium, **26, 60-61**

  and acid rain, 19

  and aging, 205

  and alfalfa, 5, 60, 61

  and behavioral problems, 178, 179

  and calcium, 43, 61, 179, 189

  and choosing basic supplement, 221

  and complete feeds, 10

  deficiencies of, 61

  and diarrhea, 189

  and electrolytes, 164

  and endurance, 61, 164

  and energy, 60

  and exercise, 61

  functions of, 60

  in grains, 60

  and grass, 61

  in growing horses, 26

  in hays, 60, 61

  indications of needed supplements of, 61

  interactions with, 61

  and lung bleeding, 200

  and Lasix, 163

  in mares, 26

  and mineral levels, 18

  and muscles, 26, 60, 61, 164

  and potassium, 71

  and racing horses, 147

  sources of, 60

  in stock feed, 11

  supplementation of, 60, 61

  testing for, 61

  toxicity of, 61

  and tying-up/muscle pain, 211

  and vitamin B, 82

Magnesium sulfate, 163

Maintenance and alfalfa, 5

  and alfalfa-oats diet, 7

  and basic diet, 2

  and complete feeds, 13

  and essential fatty acids, 112

  and mixed hay diet, 9

  and mixed hay-oats diet, 8

  and pasture grass, 3

  and protein-sweet feed, 17

  and sweet feeds, 14

  and timothy hay diet, 4

  and timothy hay-oats diet, 6

  and vitamin B12, 49

  and vitamin E, 91, 173

  and weight, 214

  and zinc, 95

Malto dextrans, 154

Manes, 57. *See also* Hair

Manganese, **62-63**

  and aging, 204, 205

  and alfalfa, 5, 62, 63

  and arthritis, 176

  and bowed tendon/tendonitis, 181

  and brittle/cracking feet, 183

  and chondroitin sulfate, 102

  and choosing basic supplement, 221

  and complete feeds, 10

  deficiencies of, 63

  and desmitis/suspensories/curbs, 187

  and exercise, 63

  and founder/laminitis, 193

  functions of, 63

  and glucosamine hydrochloride, 119

  in grains, 63

  in grass, 62, 63

  in growing horses, 26

  in hays, 62, 63

  indications of needed supplementation of, 63

  interactions with, 63

  and iron, 63

  in mares, 26

  and mineral levels, 18

  NRC levels for, 26

  in oats, 62, 63

  and OCD, 206

  and perna mussel, 136

and protein-sweet feed, 17
and SOD, 140
sources of, 62
supplementation of, 62, 63
and timothy hay diet, 4
toxicity of, 63
and vitamin C, 87
Mares behavioral problems of, **178-79**
bones and joints in, 26
consequences of inadequate nutrition in, 23, 25
and hormones, 179
infertility in, 198
lactating, 26, 35-36, 43, 69, 91
minerals for, 26-27, 46
sample diets for, 27
vitamins for, 26-27, 73
see also Pregnant mares
Mashes, 205
MCTs (medium chain triglycerides), 165
Medium chain triglycerides, 165
Melanomas, 121, 166
Menandiones, 92-93
Menaquinones, 92
Metabolic acidosis, 153
Metabolism and carbohydrates, 147, 154
and carnitine, 156
and determining supplement needs, 217
and DMG, 162
and fat, 115, 145
and gamma oryzanol, 117
and HMB, 98
and inosine, 167
and lipoic acid, 131
and manganese, 63
and methionine, 132
and MSM, 134
and phosphorus, 69
and special diets, 147
and sulfur, 81
for three-day event horse, 148
and tying-up/muscle pain, 211, 212
and vitamin B, 40, 65, 67, 83
and vitamin D, 89
and weight, 202
and zinc, 95
Methionine, **132-33**
and aging, 204
and calcium, 133
and choosing basic supplement, 221

and cystine, 110, 132-133
deficiencies of, 133
and DMG/TMG, 111
and energy, 132, 133
and exercise, 133
functions of, 41, 132, 133, 181, 183, 187, 191, 193, 197, 209
and glucosamine, 118
indications for needed supplements of, 133
interactions with, 133
lack of information about, 59, 109, 124, 128, 132, 142
and lysine, 59
and protein, 132
sources of, 132
and sulfur, 81, 132
supplementation of, 132
and taurine, 132, 133
toxicity of, 133
and weight, 202
Methylsulfonylmethane (MSM), 118, **134-35**, 209
Milk protein, 25, 59
Milk tetany, 26
Mineral oil, 186
Minerals and aging, 204
and alfalfa-oats diet, 7
and bones and joints, 25
and consequences of inadequate nutrition, 23
determining and selecting supplements of, 217-18, 219-20
feed levels of, 18-22
for growing horses, 26-27
hair analysis for, 21, 218
in hays, 1-2
and iron, 57
for mares, 26-27, 34, 35
and mixed hay diet, 9
and mixed hay-oats diet, 8
in stock feed, 11
in sweet feeds, 10
and timothy hay-oats diet, 6
total daily need of, 37
and weight, 202
see also specific mineral
Mixed hays, 8, 9, 10, 217
Moderate work, 2, 7, 8, 79, 146, 148
Molasses, 10, 11, 46, 80, 161, 196
Molybdenum, 18, **20**
Moods, 137
Moonblindness, 121, 166
Mouth, 65

MSM (Methylsulfonylmethane), 118, **134-35**, 209
Muscles aching, 77, 87, 95, 99, 105, 129
and amino acids, 125, 128, 129, 142, 143, 153
breakdown/wasting of, 153
contraction of, 70
and electrolytes, 164
and endurance, 147, 148
fuels for, 145-46, 156
and Lasix, 163
mass/strength of, 168
pain in, **211-12**
soreness of, 153, 212
for three-day event horse, 148
and training, 147
and use requirements, 146
viral infections in, 213
see also Cramping; Tying-up
Myoglobin, 105
Myositis, 121, 166

Nasal discharge, 208, 213
National Research Council (NRC), 1, 2, 4, 9, 10, 12, 14, 15, 37. *See also specific nutrient*
Nausea, 127, 168
Nervous system, **178-79**
and calcium, 42
and fluorine, 51
and magnesium, 26, 60
and potassium, 70
and selenium, 26
and sodium, 79
and vitamin B, 48, 65, 66, 73, 83
and weight, 214
and zinc, 95
Niacin, 40, **64-65**, 73, 75, 202, 204
Nickel, 19
Night vision, 85
Nitrate ion concentrations, **22**
Norepinephrine, 137
Nutritional secondary hyperparathyroidism, 69

Oats and aging, 205
alfalfa mixed with, 7
and basic diet, 1
deficiency of, 6, 7, 8
diet discussion of, 6, 7, 8
and mineral levels, 18
mixed hays and, 8
in stock feed, 11
timothy hay and, 6

for two year olds, 32, 33
for yearlings, 31
see also specific nutrient
Obesity, **202-3**
Oils, 91, 196. *See also specific oil*
Old age. *See* Aging
Ophthalmia, 75
Orchard grass, 43
Osteochondrosis dessicans (OCD),
    206-7
    and bioflavinoids, 206
    and chondroitin sulfate, 175, 206
    as consequence of inadequate
        nutrition, 25
    and copper, 19, 46, 47, 206
    and flax oil, 206
    in foals, 19, 34
    and gamma oryzanol, 117
    and glucosamine, 175, 206
    and hemp oil, 206
    and manganese, 206
    and perna mussel, 206
    in pregnant mares, 26
    supplements for, 175
    and vitamin C, 175, 206
    and zinc, 206

Pain, **211-12**
    and bioflavinoids, 99
    and coenzyme Q$_{10}$, 105, 106
    and iodine, 55
    and lipoic acid, 130-131
    and magnesium, 61
    and phenylalanine, 137
    and selenium, 77
    and vitamin B, 73, 82
    and vitamin C, 86-87
    and vitamin E, 91
    and zinc, 95
    see also type of pain
Palm oil, 91
Palosein, 140, 141
Pantothenate, 202, 204, 212
Pantothenic acid, **66-67**
Parasites, 134, 135, 138, 185, 186,
    214-215, 217
Pasture, 3, 26, 45, 84, 85, 91, 112,
    134, 180. *See also* Grass
Performance and amino acids, 59,
    109, 124, 125, 132, 142
    and coenzyme Q$_{10}$, 105
    and complete feeds, 13
    and creatine, 160, 161
    and electrolytes, 163
    and fat, 115, 145, 146, 165

and gamma oryzanol, 117
and inosine, 122, 123, 167
and iron, 57
and lipoic acid, 131, 169
and manganese, 63
and phosphorus, 69
and protein-sweet feed, 17
and selenium, 77
and sodium phosphate, 171
special diets for, 147-48
and supplements, 147-48
and use requirements, 146, 149
and vitamin B, 53, 73, 75
and zinc, 95
see also Competition
Periodontal disease. *See* Gum
    disease
Perna mussel, **136**, 176, 206
PH, 21, 79, 151
Phenylalanine, 59, 109, 124, 128,
    132, **137**, 142
Phenylbuzatone, 163
Phosphocreatine, 69
Phosphorus, **26**, **68-69**
    and aging, 204, 205
    and alfalfa, 5, 68
    in ATP, 160
    and calcium, 26, 43, 69
    and choosing basic supplement,
        221
    and creatine, 160
    deficiencies of, 69
    and endurance, 69, 160
    and energy, 69
    and exercise, 69
    in fertilizers, 19
    functions of, 69
    in grains, 68, 69
    in grass, 68
    for growing horses, 26
    and guaranteed analysis, 10
    in hays, 4, 9, 68, 69
    indications of needed supplemen-
        tation of, 69
    interactions with, 69
    for mares, 26, 34
    and mineral levels, 19, 20
    and mixed hay diet, 9
    in oats, 68
    sources of, 68
    in stock feed, 11
    supplementation of, 68, 69
    in sweet feeds, 15
    and timothy, 4
    toxicity of, 69

see also Sodium Phosphate
Phylloquinones, 92, 93
Pigs, 132
Pituitary tumors, 104
Plants, 18, 88, 92, 99, 134, 145, 152
Pneumonia, **208**, 213
Polyphenols, 120, 166
Polysaccharide, 204
Polysaccharide storage disease, 114
Potassium, 19, **70-71**, 163, 164,
    189, 200, 211, 222
Potassium chloride, 163, 164
Potassium citrate, 159
Poultry feed, 11
Prednisolone, 197, 210
Prednisone, 197, 210
Pregnant mares and calcium, 43
    consequences of inadequate
        nutrition in, 23
    and copper, 47
    and gamma oryzanol, 116
    and iodine, 55
    in last trimester, 34
    and manganese, 63
    in OCD, 26
    and phosphorus, 69
    role of major nutrients for, 25
    and sweet feeds, 14
    and vitamin B, 67
    and vitamin E, 91
    weight of, 25
Price, of stock feed, 11
PRO-BURST, 211
Proanthocyanidin, 120, 166
Probiotics, **138-39**, 185, 189, 214
Progesterone, 179
Prostaglandins, 114
Protein, **25**, **145-46**, **170**
    and aging, 204, 205
    in alfalfa, 5, 109, 124, 128, 132,
        142
    and amino acids, 109
    and behavioral problems, 178
    and bones and joints, 25
    and bowed tendon/tendonitis, 181
    and brittle/cracking feet, 183
    and calcium, 43
    and calories, 146
    as cause of health problems, 25,
        145-46
    and consequences of inadequate
        nutrition, 23
    deficiencies of, 59
    and desmitis/suspensories/curbs,
        187

and endurance, 146
and epiphysitis, 192
and exercise, 145-46, 170
and founder/laminitis, 193
functions of, 59, 109, 124, 128, 132, 142, 145-46
in grains, 109, 124, 128, 132, 142
in grass, 109, 124, 128, 132, 142
and growing horses, 25
and guaranteed analysis, 10
in hays, 1-2, 109, 124, 128, 132, 142
and heel scratches, 197
and HMB, 98
indications/rationale for, 170
and infections, 209
and iodine, 55
and L-arginine, 168
and lysine, 58
method/timing for, 170
and mixed hay diet, 9
and MSM, 135
and muscles, 145, 170
myths about, 25, 145-46
and performance, 145-46, 170
for pregnant/lactating mares, 25, 34, 35, 36
and racing horses, 147
and skin, 191, 209
sources of, 58, 109, 124, 128, 132, 142
in stock feed, 11
and sulfur, 81
supplementation of, 59, 124
in sweet feeds, 10, 17
for three-day event horse, 148
and tying-up/muscle pain, 211
and use requirements, 146
and vitamin A, 85
and vitamin B6, 110
and vitamin B, 26, 53, 65, 73, 83
see also Amino acids; specific amino acid
PSGAGs (polysulfated glycosaminoglycans), 176
Psyllium, 186
Pyridoxine. See Vitamin B6
Pyruvate, 82, 124, 128, 142, 156

Quarter horses, 156
Quercitin, 99
Quest, 214

Race horses age of, 204
and alfalfa, 147

and amino acids, 153
basic diet for, 147
and bicarbonates, 151
and carbohydrate loading, 154-55
and carnitine, 156
and chromium, 104
and creatine, 160-61
and electrolytes, 163, 164
and fat, 165
and grains, 212
and grass, 147
and hays, 147
and HMB, 98
and lipoic acid, 131
and magnesium, 147
and phosphorus, 69
prerace treatment of, 147
and protein, 146, 147, 170
and sodium chloride, 79
and tying-up/muscle pain, 212
use requirements for, 146
and vitamin B, 73, 75
and vitamin C, 87
and vitamin E, 173
see also Speed
Red blood cells, 48, 53, 56, 57, 59, 93
Reproductive system, 85, 198-99. See also Fertility
Respiratory system, 85, 87, 109, 196, **213**
Riboflavin, **74-75**, 202, 204
Ribose. See Inosine
Rice bran, 76, 116
Runny nose, 47, 95
Rutin, 99

S-adenosylmethionine, 102, 118, 119
Salmonella, 138
Salt, 1, 4, 79, 164, 200. See also Salt block
Salt block, 6, 7, 8, 13, 55, 79, 163, 222
SAM (S-adenosylmethionine), 111
Sedatives/depressives, 61
Seed meals, 60
Seizures, 53, 73, 105
Selenium, **18**, **26**, **76-77**, **173**
and aging, 204, 205
and alfalfa, 5, 76
and bowed tendon/tendonitis, 181
and brittle/cracking feet, 183
and choosing basic supplement, 221

and coenzyme Q$_{10}$, 106
and complete feeds, 10
deficiency of, 173
and desmitis/suspensories/curbs, 187
and exercise, 77
in foals, 26
and founder/laminitis, 193
functions of, 18-19, 76-77, 173
in grains, 76
in growing horses, 26
and guaranteed analysis, 10
in hays, 76
and heaves/COPD, 196
and immune system, 26, 77
and infections, 77, 209, 213
and infertility, 198
and iodine, 55
and lung bleeding, 200
in mares, 26
and muscles, 26, 77, 173, 211
NRC recommendation for, 77
and performance, 173
and pneumonia, 208
and protein-sweet feed, 17
soil deficiency of, 18
sources of, 76
in stock feed, 11
supplementation recommendation for, 76
and thyroid, 95
and timothy hay diet, 4
and tying-up/muscle pain, 19-20, 211
and viral respiratory infections, 213
and vitamin C, 77, 173
and vitamin E, 77, 91, 173
Selenocystine, 77, 173
Selenomethionine, 77, 173
Serotonin, 178
Sex hormones, 112, 204
Shoeing, 177, 183-84
Skin burning/itching, 65
and copper, 46-47
and cystine, 109
and DMG, 111
dry, **191**
and electrolytes, 163
and essential fatty acids, 112, 113
infections of, **209-10**
and methionine, 132, 133
and niacin, 65
and vitamin A, 85
and zinc, 95, 175

SOD (superoxide dismutase), 95, **140**
Sodium, **78-79**, 221
Sodium chloride, 79, 155, 164, 186
Sodium iodide, 210
Sodium phosphate, **171**
Sodium selenite, 77
Soils, 18-19, 20, 54, 76, 77
Sore throats, 47, 95
Sorghum, 76
Soy, 189
Soy oil, 214
Soybean meal, 10, 11, 25, 76, 132, 170
Soybeans, 59, 62, 180
Special diets, for performance, 147-48
Speed, 107, 108, 115, 123, 145, 147, 156, 160. *See also* Race horses
Sprains, 47, 77, 91, 95, 99, 106, 131
Staggering, 77
Standardbreds, 156
Starch, 145
Steroids, 67, 87
Stock feed, 11
Strains, 47, 77, 91, 95, 99, 106, 131
Stress and amino acids, 59, 129, 133, 143
  and antioxidants, 131
  and bioflavinoids, 99, 100, 152
  and coenzyme Q_{10}, 106
  and copper, 47
  and cystine, 109
  and epiphysitis, 192
  and gamma oryzanol, 116, 117
  and growing horses, 25
  and infertility, 199
  and iodine, 55
  and isoleucine, 125
  and lipoic acid, 131
  and lysine, 59
  and methionine, 133
  and MSM, 134, 135
  and probiotics, 138
  role of major nutrients in, 25
  and selenium, 77
  and sodium chloride, 79
  and vitamin B, 41, 49, 53, 65, 67, 73, 75, 83, 150
  and vitamin C, 87
  and vitamin E, 91
  and zinc, 95
Sucrose, 155
Sugar, 63, 107, 145, 154. *See also* Blood sugar; Glucose

Sulfate ion concentrations, **22**
Sulfates, 63, 204. *See also* Chondroitin sulfate; Glucosamine sulfate
Sulfur, 40, **80-81**, 132, 134, 222
Superoxide dismutase (SOD), 95, **140**
Supplements and aging, 204
  and behavioral problems, 179
  by use requirements, 149
  consumer's guide to, 216
  and determining needs, 217-18
  and fertility, 198
  and performance, 147-48
  for pregnant mares, 34
  selecting, 219-222
  and sweet diet, 14
  for two year olds, 33
Suspensories, **187-88**
Sweating, 71, 77, 79, 163, 164
Sweet feeds, 10, 14-16, 17, 46, 154
Swelling, 83, 86-87, 95, 99, 105, 130, 181, 182, 187, 192, 206

Tails, 57. *See also* Hair
Taurine, 132, 133
Teeth, 42, 50, 51
Tendonitis, 113, 121, 166, **181-82**
Tendons/ligaments and chondroitin sulfate, 101, 102
  consequences of inadequate nutrition on, 25
  and copper, 46, 47
  and desmitis/suspensories/curbs, 187-88
  and glucosamine hydrochloride, 119
  and methionine, 132, 133
  and MSM, 134, 135
  and perna mussel, 136
  and sulfur, 81
  and vitamin C, 87, 172
  weakness in, 81
  see also Tendonitis
Thiamine, 26, 64, **82-83**, 150, 178, 202, 214
Thirst, 79
Thoroughbreds, 146, 156
Three-day event horse, 148
Threonine, 59, 109, 124, 128, 132, 142, 181, 183, 187, 193, 204
Throat, *See also* Sore throat
Thumps, 43, 61, 71
Thyroid, 25, 26, 54, 55, 75, 77, 95, 184, 203, 212, 215

Timothy and basic diet, 1
  and complete feeds, 10
  deficiency of, 4, 6
  diet discussion of, 4, 6
  in lactating mare diets, 35, 36
  and mineral levels, 18
  in mixed hays, 9
  oats mixed with, 6
  and protein-sweet feed, 17
  for two year olds, 32, 33
TMG (trimethylglycine), **111**, 162
Tongue, 65
TPC Labs (The Pillsbury Company), 10
Trace minerals, **26**
  see also specific mineral
Training and amino acids, 153
  basic diet for, 147
  and behavioral problems, 178, 179
  and chromium, 104
  and creatine, 160-61
  and endurance, 147
  and fat, 147, 165
  and folic acid, 53
  and HMB, 98
  and L-arginine, 127, 168
  and muscle, 147
  and protein, 146
  and TMG, 162
  two year old diet for, 33
  and weight, 147
Trembling, 43, 71, 79
Triglycerides, 104, 165
Trimethylglycine (TMG), **111**, 162
Tryptophan, 59, 64, 73, 75, 109, 124, 128, 132, 142
Tumors, 120, 121, 166, 185
Two year olds, 24, 32-33
Tying-up, **211-12**
  and amino acids, 124, 125, 128, 129, 142, 143, 153
  causes of, 55, 61, 212
  and coenzyme Q_{10}, 106
  and DMG, 111, 162
  and iodine, 55
  and isoleucine, 124
  and lipoic acid, 131
  and magnesium, 61
  and potassium, 71
  and selenium, 77
  and TMG, 162
  and vitamin B, 73, 83, 150
  and vitamin E, 91

Uckele Animal Health and Nutrition
  Laboratories, 21
Ulcers, 65, 134
Urine, 105, 211
Uveitis, 121, 166

Valine, 59, 109, 124, 128, 132, **142-43**, 153, 211
Vegetable fat, 165
Vegetable oils, 26, 90, 91, 105, 112,
  114, 115, 145, 146, 148, 202,
  214
Viral infections, **213**
  and bioflavinoids, 99, 213
  and coenzyme Q$_{10}$, 105
  and copper, 47
  and diarrhea, 189, 190
  and grape seed extract, 213
  in heart, 213
  and L-arginine, 127, 168
  and L-lysine, 213
  in muscles, 213
  pathology discussion about, 213
  and pneumonia, 208
  problem description of, 213
  and selenium, 213
  supplement program for, 213
  symptoms of, 213
  treatment notes for, 213
  and vitamin A, 213
  and vitamin C, 86-87, 172, 175,
    213
  and vitamin E, 213
  and zinc, 94-95, 174, 213
Vitamin A, **26, 84-85**
  and aging, 204, 205
  and alfalfa, 26, 84, 85
  and bones and joints, 25
  and choosing basic supplement,
    221
  and consequences of inadequate
    nutrition, 25
  deficiencies of, 85
  and diarrhea, 189
  and exercise, 85
  functions of, 85
  and gingivitis, 195
  in grains, 84
  in grass, 84, 85
  in growing horses, 26
  and guaranteed analysis, 10
  in hays, 84, 85
  and heel scratches, 197
  indications of needed supplements
    of, 85

and infections, 85, 209, 210, 213
and infertility, 198
interactions with, 85
in mares, 26
and pasture, 26
and skin, 191, 209, 210
sources of, 84
supplementation of, 26, 85
and sweet feeds, 14
toxicity of, 85
and vitamin C, 87
and vitamin E, 85
and weight, 202
and zinc, 85
Vitamin B, **26-27, 150**
  in alfalfa, 74
  and behavioral problems, 179
  and bowed tendon/tendonitis, 181
  and brittle/cracking feet, 183
  and calcium, 82
  deficiencies of, 41, 49, 52, 64, 66,
    67, 73, 74, 75, 82, 83
  and desmitis/suspensories/curbs,
    187
  and determining supplement
    needs, 217
  and diarrhea, 189
  and energy, 65, 66, 73, 75
  and exercise, 49, 53, 65, 66, 73,
    75, 82, 150
  and founder/laminitis, 193
  functions of, 40, 48, 52, 64, 66,
    67, 73, 74, 75, 82, 83,
    150, 212
  in grains, 48, 52, 64, 65, 66, 72,
    74, 75, 82, 150
  in grass, 52, 53, 72, 74, 75
  in growing horses, 26-27
  in hays, 40, 52, 72, 74, 75
  indications for needed supplemen-
    tation of, 41, 49, 52,
    64, 66, 67, 73, 74, 75, 82, 83
  interactions with, 41, 49, 52, 64,
    66, 67, 73, 74, 75,
    82, 83
  in mares, 26-27
  method/timing of, 150
  in oats, 52, 72, 74, 75
  and performance, 150
  sources of, 40, 48, 52, 64, 66, 72,
    74, 82, 150
  supplementation of, 26-27, 41,
    49, 52, 64, 66, 67, 73,
    74, 75, 82, 83
  toxicity of, 41, 49, 52, 64, 66, 67,

73, 74, 75, 82, 83
and weight, 202, 214
see also specific vitamin
Vitamin B6, **72-73**
  and aging, 204
  and behavioral problems, 178
  and bowed tendon/tendonitis, 181
  and brittle/cracking feet, 183
  deficiences in, 73
  and desmitis/suspensories/curbs,
    187
  and founder/laminitis, 193
  functions of, 73
  and gamma oryzanol, 117
  and heel scratches, 197
  and HMB, 98
  indications of needed supplemen-
    tation of, 73
  and infections, 209
  interactions with, 73
  and methionine, 133
  and MSM, 135
  and phenylalanine, 137
  and protein, 110
  and riboflavin, 75
  and skin, 191, 209
  sources of, 72
  supplementation of, 72, 73
  toxicity of, 73
  and tying-up/muscle pain, 212
  and weight, 202
Vitamin B12, 18, 44, 45, **48-49**, 52,
  87, 150, 172
Vitamin B15, **111**. *See also* DMG
Vitamin C, **26, 86-87, 172**
  and aging, 204, 205
  as antioxidant, 120, 130
  and arthritis, 176
  and bioflavinoids, 87, 99, 100,
    152, 172, 176
  and bowed tendon/tendonitis, 181
  and brittle/cracking feet, 183
  and chondroitin sulfate, 102
  and choosing basic supplement,
    221
  and connective tissues, 26
  and copper, 87
  deficiencies of, 86
  and desmitis/suspensories/curbs,
    187
  and diarrhea, 189
  and energy, 172
  and exercise, 87, 172
  and founder/laminitis, 193
  functions of, 86, 130, 152, 172, 175

and gingivitis, 195
and glucosamine hydrochloride, 119
and grape seed extract, 120, 166
in grass, 86
in growing horses, 26
and hay, 26
and heaves/COPD, 196
and immune system, 26, 86-87
and infections, 86-87, 172, 175, 213
and infertility, 198
and iron, 57, 87
and lipoic acid, 130-131
and lung bleeding, 200
and manganese, 87
in mares, 26
and MSM, 135
and muscles, 87, 172
and OCD, 175, 206
and performance, 172
and perna mussel, 136
and phenylalanine, 137
and pneumonia, 208
and selenium, 77, 173
and skin infections, 209
sources of, 86
supplementation of, 26, 86
toxicity of, 172
and tying-up/muscle pain, 211
and vitamin A, 87
and vitamin B12, 172
and vitamin E, 87, 91, 172, 173
and weight, 202
and zinc, 87
Vitamin D, **26**, **88-89**
and aging, 204
and bones and joints, 25
and calcium, 43, 88, 89
and choosing basic supplement, 221
and complete feeds, 10
and consequences of inadequate nutrition, 23
deficiencies of, 89
functions of, 88
in growing horses, 26
indications for needed supplementation of, 89
interactions with, 89
for mares, 26
sources of, 88
supplementation of, 88, 89
and sweet feeds, 14
toxicity of, 89

Vitamin E, **90-91**, **173**
and aging, 204, 205
and alfalfa, 5
and alfalfa-oats diet, 7
as antioxidant, 120, 130
and choosing basic supplement, 221
and coenzyme Q$_{10}$, 106
and diarrhea, 189
and endurance, 173
and exercise, 91
function of, 130
in grains, 90
in grass, 90, 91
in hays, 90
and heaves/COPD, 196
and heel scratches, 197
and immune system, 91
and infections, 91, 209, 210
and iron, 57, 91
and lipoic acid, 130-131
and lung bleeding, 200
and mixed hay diet, 9
and mixed hay-oats diet, 8
and muscles, 91, 211
and performance, 173
and pneumonia, 208
and protein-sweet feed, 17
and selenium, 77, 91, 173
in stock feed, 11
and timothy hay diet, 4
and timothy hay-oats diet, 6
and tying-up/muscle pain, 211
and viral respiratory infections, 213
and vitamin A, 85
and vitamin C, 87, 91, 172, 173
and weight, 202
and zinc, 91
Vitamin K, 75, **92-93**, 200, 222
Vitamins and consequences of inadequate nutrition, 23
freshness of, 221
for growing horses, 26-27
in hays, 1-2
human, 221
in lactating mare diets, 35
for mares, 26-27
total daily need of, 37
and weight, 214
see also Supplements; *specific vitamin*

Warfarin, 93
Water, 79, 107, 163, 164, 171, 186,
200
Water soluble vitamins. *See* Folic acid; Niacin; Pantothenic acid; Riboflavin; Thiamine; Vitamin B6; Vitamin B12
Weakness, 71, 79, 89
Weaning, 36
Weanlings, 23, 24, 25, 28-29
Weather, 79, 85
Weight and behavioral problems, 178
and calories, 214
and endurance, 147
and epiphysitis, 192
and fat, 114, 115, 165
and feeds, 214-15
of foals, 25
gain of, 192, **202-3**
and grains, 214
and hard work, 214
and iodine, 55
loss of, 180, 190, **214-15**
and lysine, 202
and maintenance, 214
and minerals, 202
and nervous system, 214
of pregnant mares, 25
and probiotics, 138, 139
and protein, 145-46
and training, 147
and use requirements, 146, 214-15
and vitamin A, 202
and vitamin E, 202
and vitamins, 214
Weights and measures, 38, 219
Wheat bran, 76, 94
Whey, 59
White blood cells, 56, 91
White muscle disease, 19, 26
Wild horses, 3
Winstrol, 116
Wood chewing, 13
Work level, 12, 214-15, 217. *See also specific level*
Working horses, 6, 8, 83, 91. *See also* Heavy work; Light work; Moderate work
Worming, 138, 185, 214
Wounds and antioxidants, 131
and bioflavinoids, 99, 100, 152
and coenzyme Q$_{10}$, 106
and cystine, 109, 110
and leucine, 128
and lipoic acid, 131

and selenium, 77
and vitamin E, 91
and zinc, 95

Yearlings, 23, 24, 25, 30-31
Yeast and aging, 204
   and bloating/gas, 180
   and pantothenic acid, 66
   and probiotics, 139
   vitamin B in, 48, 52, 64, 72, 74,
     82, 150
   and weight, 214
Yogurt, 189

Zinc, **20, 94-95**
   and aging, 204, 205
   and alfalfa, 5
   and arthritis, 176
   and bowed tendon/tendonitis, 181
   and brittle/cracking feet, 183
   and choosing basic supplement,
     221

and complete feeds, 10
and copper, 47, 95
and desmitis/suspensories/curbs,
   187
and determining supplement
   needs, 217
and exercise, 95
and founder/laminitis, 193
functions of, 94-95, 174-75
and gingivitis, 195
in grains, 94
in growing horses, 26
in hays, 94
and heaves/COPD, 196
and heel scratches, 197
and hooves, 41
and immune system, 95, 174, 176
indications for needed supplementation of, 175
and infections, 95, 174, 209, 213
and infertility, 198
and iodine, 55

and iron, 57
and lung bleeding, 200
for mares, 26, 34
and mineral levels, 19, 20
and muscles, 95, 211
NRC levels for, 26
and OCD, 206
and pneumonia, 208
and protein-sweet feed, 17
and respiratory system, 213
and skin, 191, 209
and SOD, 140
soils deficient in, 19
sources of, 94
supplementation of, 94, 175
and timothy hay diet, 4
and tying-up/muscle pain, 211
and vitamin A, 85
and vitamin C, 87
and vitamin E, 91
for weanlings, 28, 29
for yearlings, 30, 31